THROUGH THE LOOKING GLASS

A CITIZEN'S DO-IT-YOURSELF GUIDE TO CLIMATE SCIENCE

PAUL MACRAE

To climate realists

everywhere,

with special thanks to Patrick Hunt,

and to my wife Sheila,

who always helps keep me real

Climate Realists of British Columbia

https://climaterealists.ca

Climate Realists of British Columbia

The 'Science' of climate change is 'settled'—or is it?

In the sense that it [climate change] will be an urgent catastrophe which will eliminate all life, that kind of statement is absolutely unwarranted.

—Climate scientist Stephen Schneider

In climate research and modeling, we should recognize that we are dealing with a coupled non-linear chaotic system, and therefore that long-term prediction of future climate states is not possible.

—IPCC 2001 report

We need to get some broad-based support, to capture the public's imagination. That, of course, entails getting loads of media coverage. So we have to offer up scary scenarios, make simplified, dramatic statements, and make little mention of any doubts we might have.

—Stephen Schneider

The data don't matter.... We're not basing our recommendations on the data. We're basing them on the climate models.

—Climate scientist Chris Folland

Uncertainties are so endemic in these complex [climate] problems that suffer from missing data, incomplete theory and nonlinear interactions.

—Stephen Schneider

Government and UN press releases and summaries do not accurately reflect the [IPCC] reports themselves. ... Distinguished climate experts (including report authors themselves) are embarrassed by some media portrayals of the science.

—Steven Koonin, *Unsettled*

Review by experts and governments is an essential part of the IPCC process.

—IPCC mission statement

The scientific community would come down on me in no uncertain terms if I said the world had cooled from 1998. OK it has but it is only seven years of data and it isn't statistically significant.

—Phil Jones, head of the Climatic Research Unit

The problem is that in areas of science which are seeking to understand the behavior of large complex systems which can't be replicated in the lab, it is very hard if not impossible to apply the scientific litmus test of falsification through experimentation. ... Climate scientists frequently have to reach their conclusions on the basis of the partial, and sometimes poorly tested, evidence and models available to them.

—Mike Hulme, climate scientist

Ultimately, there are three core values in western society, and for that matter, in global society, that have to change. One core value is about growth and materialism. The second core value is liberty and individualism, which has to be rethought because the kind of individualism that is preached by neo-liberals is part of the problem.

—Public Health Agency of Canada report

I tried hard to balance the needs of the science and the IPCC, which were not always the same. I worried that you might think I gave the impression of not supporting you well enough while trying to report on the issues and uncertainties.

—Climate scientist Keith Briffa to Michael Mann

The fact is that we cannot account for the lack of warming at the moment and it's a travesty that we can't

—Climate scientist Kevin Trenberth

Alarmist: A person who needlessly alarms or attempts to alarm others, as by inventing or spreading false or exaggerated rumors of impending danger or catastrophe.

—American Heritage Dictionary

Contents

Introduction: Through the Looking Glass

It was all very well to say "Drink me," but the wise little Alice was not going to do that in a hurry. "No, I'll look first," she said, "and see whether it's marked 'poison' or not."

This book is not for the faint of heart. For more than thirty years we have been told that the planet is warming dangerously, even apocalyptically; that the reason for this dangerous warming is increased carbon dioxide from our fossil-fuel use; and that we're all gonna die if we don't cut back on our carbon-spewing ways by *severely* reducing our standard of living.

Furthermore, this apocalyptic viewpoint is, we are told, backed by solid Science (a.k.a., "The Science" or "The Scientific Consensus"); if you don't subscribe to this view, you are a "climate denier" and therefore a very bad person, on a moral level with Holocaust deniers.

This book argues that almost none of the above is true. But it takes courage and an act of faith to go through the "looking glass" from the widely held and intensely promoted "scientific consensus" that we are facing a global-warming catastrophe to a view that is much more skeptical of these claims.

To better understand these two views of our climate future, we're going to go, like Lewis Carroll's Alice, though the looking glass, in this case the climate looking glass (or, if you prefer, down the climate-science rabbit hole).

Alice in Wonderland and *Through the Looking Glass* are both great fun as Carroll (a.k.a., Charles Dodgson) parodies the absurd logic and sheer insanity that Alice encounters on the "other" side of the mirror, where the White Queen can believe "six impossible things before breakfast."

Carroll's broader target for social satire was, of course, the illogical and sometimes absurd thinking of his own society and time (*Alice in Wonderland* was published in 1865: the original John Tenniel illustrations of the Red Queen bear more than a passing resemblance to Queen Victoria).[1]

We're going to make a similar journey between worlds, but if you're willing to make the journey, I promise I will show you, with ample evidence, that almost everything you've been told about catastrophic, human-caused "climate change" is misleading, exaggerated, or just plain wrong—as absurd, indeed, as the insanity Alice faces in Wonderland and on the other side of her drawing room mirror.

Is carbon dioxide the 'control knob' of climate?

The cornerstone of alarmist climate science is the assertion that carbon dioxide is the "control knob" of climate.[2] This means that adding more carbon dioxide to the atmosphere will eventually lead our planet to uncontrolled warming.

The alarmist flipside is that drastically *reducing* carbon dioxide levels by radically reducing our use of fossil fuels (aiming for "Net Zero by 2050"), and replacing them with more "sustainable" energy sources, will—through a reverse "control knob" effect—bring temperatures back to livable levels.

This book argues the case for a different theory: yes, carbon dioxide is a "greenhouse gas," but carbon dioxide is *not* the climate "control knob," and never was, and that reducing our "carbon footprint" will have a virtually zero effect on climate, now or in the future.

I'll be using both direct and indirect evidence.

As *direct* evidence, I will introduce you to the *actual temperature records* of the last few decades to show you that the planet is not warming nearly as quickly or intensely as the alarmist climate scientists claim; indeed, for long periods in the past 70 years the planet has barely warmed at all. And I will give you, the reader, opportunities to test my claims by consulting the temperature data yourself—hence a "do-it-yourself" guide.

As *indirect* evidence of the weakness of the "control knob" theory I offer many examples of alarmist climate scientists and environmentalists trying to avoid, bury, or muddle evidence that would prove "The Science" is not as "settled" and "certain" as its promoters would like us—the public, the media and the politicians—to believe. I'll also offer you many cases where climate alarmists have out-and-out lied to us. If the science is "settled," why the need to lie or obfuscate?

In contrast, this book argues that not only is carbon dioxide *not* the planet's "control knob" now, but the long-term geological record shows that CO_2 is either not at all or poorly correlated with warming levels over hundreds of millions of years. When there does appear to be a correlation, temperature change almost always *precedes* CO_2-level change (my evidence for these claims is in Chapters 1-4).

If this is true, then the Net Zero theory that *reducing* CO_2 levels will reduce "global warming" cannot be true and will have no beneficial effect.

This book also argues that increased carbon emissions will not warm the planet much more than they already have, thanks to the "carbon-band-saturation effect," discussed in Chapter 6; in fact, more CO_2 is actually *good* for the planet and for humanity by encouraging enhanced plant growth (the "greening of the planet"), discussed in Chapter 8.

'Climate crisis' based on 'post-normal' science

In Chapter 11 you'll learn that the apocalyptic vision of a blasted world is not based in traditional empirically based science, although this is what alarmists would like us to believe, but on a distorted version of traditional science called "post-normal" science.

Unlike traditional science, "post-normal" science relies not on objective empirical evidence, rigorous testing and risky predictions, but on the *subjective* judgments of *politicized* climate scientists, activist environmental groups and the United Nations, to produce a "consensus" based on computer models, a "consensus" that we, the public, are not supposed to question or criticize.

This aversion to criticism is a sure sign that "The Science" is not nearly as "settled" and "certain" as we have been led to believe, and may even be dead wrong.

In Chapters 13-16, this book argues that even if catastrophic warming was true (it isn't, but if), drastically cutting back CO_2 emissions, will not "save" the planet or its people. Instead, extreme anti-carbon policies will damage or destroy an economic system based on fossil fuels and free markets that is making this planet ever more livable for billions of people—an economic system that can *help* us adapt to warming, if that occurs, if we let it.

Why use the term 'alarmist'?

It has been suggested to me that the term "alarmist" is polarizing and shouldn't be used, but I prefer this term.

The dictionary definition of "alarmist" is

> a person who needlessly alarms or attempts to alarm others, as by inventing or spreading *false or exaggerated rumors of impending danger or catastrophe.*[3] [emphasis added]

This is very much the argument of this book: while there may be climate challenges ahead, climate *alarmism* is, indeed, as I will show you with ample evidence, based on "false or exaggerated rumors of impending danger or catastrophe."

As for the term "alarmist" being polarizing: we *should* be polarized. We *should* be angry that a group of politicized scientists, environmental zealots and utopian idealists are threatening our livelihoods, lifestyles and even our very lives in the name of a "climate crisis" that is mostly a product of the alarmists' imaginations, of computer models rather than empirical evidence, and of radical ideologies rather than objective "science."

By contrast, although I used to think of myself as a climate "skeptic," I now prefer the term climate "realist"—"realist" because I favor a *realistic* approach to the evidence for and against so-called "human-caused" (anthropogenic) and supposedly catastrophic global warming, and a *realistic* approach to how we should deal with the climate change that does exist.

Realistic alternatives to Net Zero by 2050

This book's view is, then, pretty much the polar *opposite* of what the Intergovernmental Panel on Climate Change (IPCC) and its supporters tells us—the public, politicians and the media. In this book we'll go through the "looking glass" that separates the two positions—alarmist and realist—and survey the climate issue from the realist side. Then we can decide which side—climate alarmism or climate realism—makes the most political, economic, social and environmental sense.

And you, reader, are quite right to be wary when someone tells you that the *scientists* have got it all wrong, especially if the someone making this argument isn't a scientist, which I'm not (but then, neither are Greta Thunberg or Al Gore and they're accepted as authorities on climate).

In my defence, I was a journalist for 35 years with several newspapers, including Toronto's *Globe and Mail, The Bangkok Post,* and the Victoria *Times-Colonist,* and then for 15 years an instructor in writing at the University of Victoria.

So although I'm not a scientist, much less a climate scientist, I do have more experience than most people in both journalistic and academic research, and perhaps I can present the results of my research in an academically rigorous way, i.e., backed up by evidence (while also being, I hope, readable). In other words, I might have some credibility as a *lay investigator* of climate science, even if I'm not a climate scientist.

True views are 'harmonious'

The Greek philosopher Aristotle (384-322 BC) is generally credited with inventing the practice of science in the West, and one of his statements from more than 2,000 years ago resonates with me:

> With a true view all the data harmonize, but with a false one the facts soon clash.[4]

This is a beautiful and useful way of looking at science (or any discipline, including politics and economics). If a scientific theory has much evidence that "harmonizes" with it, it's probably a good theory and worth accepting.

But if a scientific theory has all sorts of evidence that doesn't "harmonize," then it's almost certainly an incoherent theory and should be dropped, or at least not accepted without considerable skeptical scrutiny. (It's striking how insistent alarmist climate science is that politicians, the media and the public *not* scrutinize its theory because, of course, "The Science is settled.")

The French thinker Jean François Revel makes this point in a different way:

> How is it possible for a theory, which is false in its component parts, to be true as a whole?[5]

How indeed? I'll show you many examples of "component parts" of alarmist climate science that are clearly false.

And, in the same vein, Thomas Huxley (1825-1895), a British Victorian who was also one of the founders of modern science (he was "Darwin's bulldog" in defending Darwin's evolutionary theory), wrote:

> The great tragedy of science—the slaying of a beautiful hypothesis by an ugly fact.

A more modern term for scientific claims that don't "harmonize," that have "false component parts" or "ugly facts," i.e., are not coherent, is "anomaly." Scientists normally change or abandon their theories when they are facing so many anomalies that the theory is no longer tenable.

In this book, I am going to turn official, alarmist climate science on its head by showing you *many* anomalies, *many* cases in climate science where the data not only doesn't "harmonize" but gets downright ugly.

So I'm going to ask for an act of faith. Hear me out. Look at the facts—I emphasize the word *facts*—I present. Go to the interactive websites that I suggest and check my temperature data for yourself (again, this is a "do-it-yourself" guide). Then decide if my case makes both rational and scientific sense, regardless of what official, alarmist climate science claims.

One thing I guarantee: even if you end up disagreeing with the realist position, you will at least be somewhat more skeptical of the apocalyptic alarmist claims of climate science and climate scientists. I'd consider that a good return for my efforts.

I've tried to keep the book's argument as direct as possible, so I've put supporting material that is relevant (or just interesting, or maybe boring but worth knowing) but not essential to my main argument into a series of "Sidebars," for you to read or not, as you wish. My references are at the end of each chapter; most of the references you can check for yourself online.

I've called this a "citizen's guide" because I strongly believe that the success of a democracy is based on citizens getting *accurate* information on which to base their political decisions. With a more accurate or at least more balanced picture of the climate situation, when readers go to the polling booths I hope they will pick politicians and parties that share this realist position, rather than (as is almost universally the case now) politicians and parties with the alarmist viewpoint.

I'll begin with an example of a massive geological event—the gradual fall in temperature over the past 50 million years leading to our current Ice Age—that supports two of my basic points in this book:

1. When temperature and CO2 appear to be correlated, CO2 levels generally *follow* temperature changes; and
2. Alarmist climate scientists and their followers are so mesmerized by the CO2 "control knob" theory that they can't see contrary evidence right in front of their eyes.

Notes for Introduction

---

[1] For a delightful look at Carroll's satirical targets and esoteric meanings, see David Day, *Alice's Adventures in Wonderland Decoded: The Full Text of*

Lewis Carroll's Novel with its Many Hidden Meanings Revealed. Doubleday Canada, 2015.

[2] Andrew A. Lacis et al., "Atmospheric CO_2: Principal Control Knob Governing Earth's Temperature." *Science*, Oct. 15, 2010, pp. 356-359. Available online but often behind a pay wall, although there are repostings on some public websites. For a summary of Lacis's paper see https://pielkeclimatesci.wordpress.com/2010/11/03/guest-post-co2-the-thermostat-that-controls-earth%e2%80%99s-temperature-by-andy-lacis/.

[3] "Alarmist." *American Heritage Dictionary.*

[4] Aristotle, *Ethics*. Book 1, Section 8.

[5] Jean-François Revel, *Neither Marx nor Jesus.* New York: Laurel, 1971, p. 15

Chapter 1: 50 Million Years of Cooling

The proper order of things is often a mystery to me.

In more than fifteen years of research on climate, I have particularly enjoyed learning about paleo-history and paleo-geography, that is, the history of our planet going back thousands and millions of years. And in my reading, I have often encountered the following assertion:

Fifty-five million years ago, about 10 million years after the death of the dinosaurs, the Earth went through a strong warming called the Paleo-Eocene Thermal Maximum (PETM). The PETM was 5-8° Celsius warmer than today—or about 20-22°C on average—with a CO_2 level as high as 3,000 parts per million (ppm), compared to today's paltry 425 ppm.[1]

The PETM was likely caused by a huge volcanic eruption, followed by warming that released vast quantities of undersea methane, a powerful greenhouse gas, although its effects in the atmosphere are relatively short-lived. As the authors of the paper cited here conclude: "This [extreme PETM warming] supports the view that climate sensitivity increases substantially when greenhouse-gas concentrations are high."[2]

So far, so good; greenhouse-gas theory of global warming apparently confirmed! But then something odd happened.

Following the PETM, and despite very high levels of CO_2, about 50 million years ago the Earth began to *cool*. The planet got colder. And colder. And colder. This cooling has continued to today, with ups and downs but always trending down (see Figure 1.1).

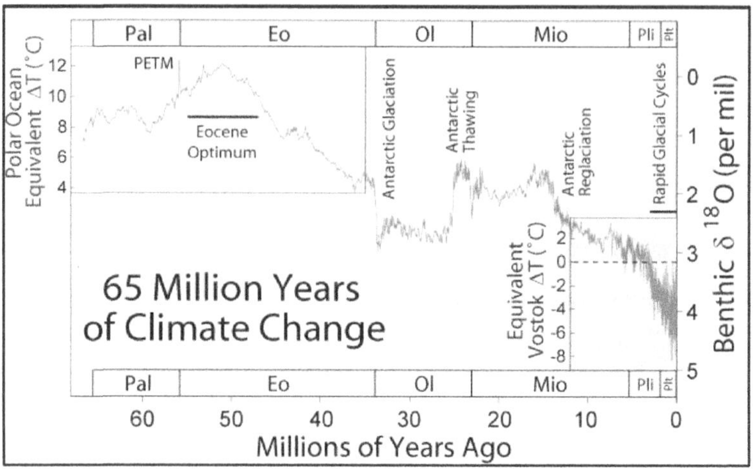

Figure 1.1: Fall of temperature over 65 million years. Source: Global Warming Art/Wikipedia

The cooling eventually became so extreme that, about 2.6 million years ago, the Earth entered an Ice Age, with glaciers advancing and retreating in the northern regions of North America, Europe and Asia (the Southern Hemisphere is mostly water and therefore less affected since glaciers form on land). During all this time CO2 levels also fell, and many writers on climate have concluded that the drop in CO2 *caused* this multi-million-year cooling.

Tectonic change, not CO_2, caused cooling

For example, geologist John Dvorak, in his excellent book *How the Mountains Grew: A Geological History of North America,* explains that a few million years after the PETM, "two tectonic events started the planet on a long and slow Cenozoic cooling." One event was the uplifting of the Himalayan Mountains about 50 million years ago as part of the massive slow-motion collision between the Indian tectonic plate and Eurasia. Dvorak writes:

> It was the rise of these mountains–and the newly exposed rock in the steep mountain cliffs–that increased the rate of physical and chemical weathering, which in turn *drew down the amount of carbon dioxide in the atmosphere and caused the planet to cool.*[3] [emphasis added]

The second tectonic event, Dvorak notes, was

> the movement of Antarctica over the South Pole and its isolation from the other continental masses. ... The isolation of Antarctica became more severe now that it was surrounded by a circumpolar sea. The warm water currents of the Pacific and Atlantic could no longer reach it, and Antarctica cooled faster than the rest of the planet. That was 35 million years ago."[4]

Finally, he notes a third "key event" about three million years ago that precipitated the current ice age: "The creation of the Isthmus of Panama and the closure of the seaway between the Pacific and Atlantic."[5]

Similarly, science writer Lewis Dartnell, in his very readable *Origins: How Earth's History Shaped Human History*, notes:

> The past 50 million years or so have been characterized by a chilling of the global climate. This process is called the Cenozoic cooling, and it culminated 2.6 million years ago in the current period of pulsing ice ages. ... This long-term cooling trend has been largely driven by the continental collision of India [the Indian tectonic plate] into Eurasia and the raising of the Himalayas [about 50 million years ago].
>
> The subsequent erosion of this towering range of rocks has scrubbed a lot of carbon dioxide out of the atmosphere, resulting in a reduction of the greenhouse effect that was previously insulating the planet ... and leading to declining temperatures. In turn, the generally cooler conditions drove less evaporation from the oceans to create a less rainy, drier world.[6] [emphasis added]

Dartnell then describes, like Dvorak, the two other tectonic factors that also cooled our world: the gradually chilling circumpolar current around Antarctica that brought permanent ice to that continent by about 14 million years ago, and the joining of South and North America at the Panama Isthmus about three million years ago.[7]

Rising of the Himalayas

So, let's unpack this a bit. About 50 million years ago, following the ultra-warm PETM (in short, a "climate crisis." Yet somehow the Earth and its animals survived this extreme heat; perhaps humans can, too), and despite very high levels of carbon dioxide, the planet began to cool. The initial cause of this cooling, as both Dvorak and Dartnell (and many others) note, was continental drift: the collision of the Indian tectonic plate into Eurasia, which created the Himalayan mountains and the Tibetan Plateau.

The Himalayas are the world's largest and highest mountain range. So, beginning 50 million years ago, air currents that used to circulate at sea level and stay warm were now forced to rise as they moved up the mountain slopes. Naturally, the air cooled and the wind that swept down *off* these forming mountains was cold. Result? Gradually cooling climate.

Also, because high mountain peaks and plateaus are colder than ground level, they can accumulate snow and ice year-round, even in tropical climates (Mt. Kilimanjaro in Tanganyika comes to mind). So as the Himalayas formed, increasing ice and snow on the range's peaks and plateaus radiated more of the sun's heat back into space. Result? Gradually cooling climate.

And the Himalayas weren't the only major, planet-cooling mountain range forming in this 50-million-year continental drift into our current Ice Age. Tectonic collisions in North America began to form the Rockies about 80 million years ago (leading to cooling).

The crash of the African plate into Europe started building the Alps about 35 million years ago (more cooling). The Andes mountains are 30 million years old (cooling). Like the Himalayas, these new mountain ranges also created colder air currents and ice and snow on their peaks reflected more of the sun's radiation back into space (cooling).

Think about it. If you put an open refrigerator in a room, the room will cool. The new mountain ranges acted like refrigerators, making the planet colder.

Sure, as Dartnell and Dvorak both note, there's newly exposed rock on the mountainsides, and water run-off will dissolve carbon dioxide in the rocks as it streams down the mountain slopes, and this process will gradually reduce CO2 levels ("chemical weathering", "erosion").

In their descriptions, Dartnell and Dvorak are channeling alarmist climate science's doctrine of CO2 as the "control knob" of climate. But surely mountain formation is the *dominant* player in this cooling scenario, not dissolved CO_2. However, alarmists are so fixed on the carbon-dioxide theory of warming that they can't see the evidence right in front of their noses.

The cooling of Antarctica

The relative unimportance of falling CO_2 levels to planetary cooling becomes clearer when we consider the second of the two tectonic events: the final breakup, about 40 million years ago, of the Pangean super-continent. In this breakup, South America and Australia drifted away from Antarctica, which created a circumpolar ocean current around the southern continent (see Figure 1.2; A.C.C. = Antarctic Circumpolar Current).

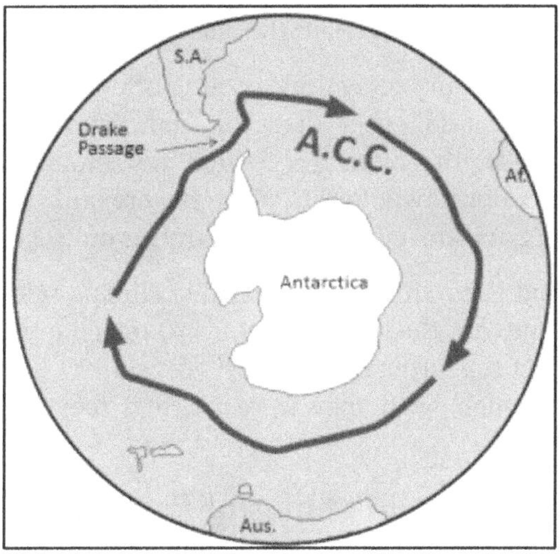

Figure 1.2: Circumpolar current caused by tectonic drift of South America and Australia gradually freezes Antarctica. Source: BCCampus, Physical Geography, 16.1 Glacial Periods in Earth's History

Before this epic tectonic event, the interchange of warm and cool currents kept Antarctica warm enough for tropical conditions 40 million years ago, and allowed plant life up to 14 million years ago. After that, Antarctica got colder and colder with each revolution of the current around the continent until Antarctica became entirely ice-bound.

Earth now had an enormous freezer in its basement that, when added to the many mountain refrigerators, meant the planet would cool. Inevitably. Regardless of the CO_2 levels.

Falling CO_2 due to "chemical weathering" and "erosion" may have played a role in this cooling, but it had to be a small role in the face of these two massive tectonic events (plus many other tectonic events that were less massive but still significant).

Why a small role for CO_2? Because in "chemical weathering," yes, water leaches CO_2 out of mountainsides and deposits it, eventually, in the oceans. But these oceans were already cooling in response to tectonic events. Cold water can hold more dissolved CO_2 than warmer water. We see this with cold cans of cola; as the liquid gets warmer, the CO_2 that gives soda its fizz gradually evaporates and the beverage goes "flat."

Similarly, over millions of years, while tectonic events cooled the planet, colder oceans soaked up carbon dioxide and CO_2 levels fell in response to the overall cooling; diminishing atmospheric CO_2 may have *contributed* to cooling, but it didn't *cause* this cooling, as claimed by Dartnell and Dvorak and many other writers who have uncritically accepted the alarmist "control knob" narrative.[8]

This theory of tectonic cooling offers a much more logical relation of cause and effect than the alarmist thesis that falling CO_2 levels *caused* this planetary cooling. It's astonishing that Dartnell and Dvorak don't twig to this, since they're both aware that mountain-building and the isolation of Antarctica jump-started the cooling process.

But then, this is what alarmist climate science teaches: levels of CO_2 rule (are the "control knob" of) the climate. Dartnell and Dvorak are just repeating the alarmist "consensus." If they had paused to think through what they were writing, they might have told a different story.

Closing of Isthmus of Panama

As noted above, the final nail in Earth's cooling coffin was the joining of the North and South American continents at the Isthmus of Panama about three million years ago (some say 2.8 million years ago[9]).

Up to then, the mingling of the Atlantic and Pacific currents had a moderating effect on global temperature. With this moderating effect removed, permanent ice caps also began to form in the Arctic, further adding to the tectonic cooling effect (a freezer in the basement *and* the attic!). And, yes, the Arctic ice caps are a relatively *new* feature in Earth's recent geological history.

At this point, Earth descended into a full-blown Ice Age, with a roller-coaster climate of glacials (cold) and interglacials (warm) to become *the coldest the planet has been in 250 million years,* and perhaps in

500 million years.[10] That's certainly something to think about when climate alarmists claim we are becoming too warm! We'll look at this glacial roller-coast in detail in Chapter 2

So, to summarize, we have had three major tectonic events in 50 million years, along with many others almost as powerful (e.g., the uplift of the Rockies and Alps), all of which tended toward *cooling* rather than warming. And, at the same time, CO2 levels fell from 3,000 ppm to below 200 ppm. But which came first: falls in CO2 levels leading to cooling, as orthodox climate science teaches? Or cooling leading to falls in CO2 levels?

Looking at the geological evidence with an open mind, the tectonic scenario—planetary cooling causing the decline in CO2 levels—makes more sense than the "dissolved CO2 from mountains leading to cooling" scenario preferred by climate alarmists.

In the more logical tectonic scenario, falls in temperature *preceded* the falls in CO2 levels of the last 50 million years. The tectonic scenario "harmonizes" with the known facts of the past 50 million years far better than the alarmist, International Panel on Climate Change (IPCC) scenario.

I've always been suspicious of the CO2-drives-temperature theory, but the case history of this 50 million years of cooling ultimately convinced me: CO2 is not the *primary* driver of climate ups and downs, although, of course it plays a role. However, this role is nowhere near as crucial as alarmist climate science would have us believe.

In Chapter 2, we'll look at another solid reason to question the orthodox climate narrative: the relation of carbon dioxide and temperature during the Ice Age of the past 2.6 million years.

Notes for Chapter 1

---

[1] Wikipedia, "Paleocene–Eocene Thermal Maximum." See also Tierney, Jessica E., et al., "Spatial patterns of climate change across the Paleocene–Eocene Thermal Maximum." *PNAS*, Vol. 119, No. 42 (Oct. 10, 2022). Available at https://doi.org/10.1073/pnas.2205326119.

[2] Wikipedia, PETM.

3 John Dvorak, *How the Mountains Grew: A Geological History of North America*. New York: Pegasus Books, 2021, p. 325.

4 Dvorak, p. 325.

5 Dvorak, p. 326.

6 Lewis Dartnell, *Origins: How Earth's History Shaped Human History*. New York: Basic Books, 2019, pp. 9-10.

7 Dartnell, pp. 42-43.

8 Edward Aguado and James E. Burt, *Understanding Weather & Climate*. Upper Saddle River, NJ: Pearson Education, 2004, p. 495. This is a university-level textbook.

9 Texas A&M news release, "New study reaffirms timeline on formation of Isthmus of Panama." 2023.

10 National Oceanographic and Atmospheric Administration (NOAA), "What's the coldest the Earth's ever been?" *Climate.gov*, Feb. 18, 2021.

Chapter 2: The two-million-year roller coaster

The Red Queen shook her head. "You may call it 'nonsense' if you like," she said, 'but I've heard nonsense, compared with which that would be as sensible as a dictionary!"

The ultra-celebrity of former U.S. vice-president Al Gore is perhaps fading these days, but in 2006 he produced and narrated a movie, *An Inconvenient Truth*, that preached alarmist climate science to the masses and even won him an Oscar and a piece of a Nobel Peace Prize (the IPCC got the other piece).

The centrepiece of Gore's movie was a massive wall chart showing the ups and downs of temperature (white bottom line) and carbon dioxide (grey top line) over 650,000 years (see Figure 2.1). The last bit of the CO2 part (extreme top-right of the chart) leaps up like a frantic puppy, so high that Gore had to stand on a lift to reach it! Scary!

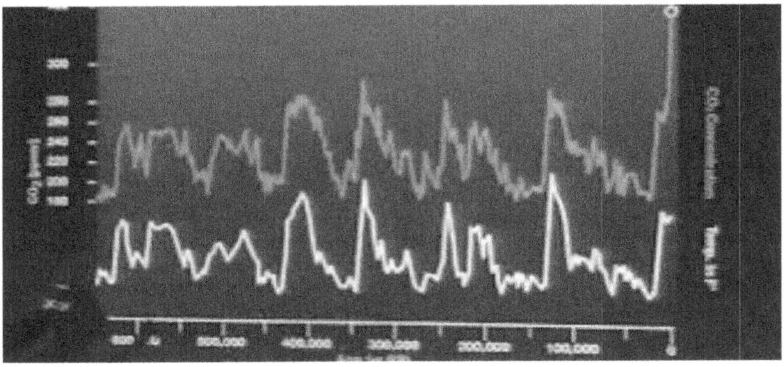

Figure 2.1: Al Gore's 650,000-year graph of temperature and carbon dioxide from An Inconvenient Truth

Gore's graph illustrates the last five or six of the series of glacials (colder) and interglacials (warmer) that have dominated Earth for the past 2.6 million years—we live in an Ice Age! In fact, as noted earlier, our current geological time is the *coldest* the planet has been in more than *250 million years,* and perhaps in 500 million years.[1]

While we're looking at Figure 2.1, note that the temperature of most recent interglacial peak, the Eemian of 125,000 years ago, is quite a bit *higher* than today's temperature peak so far (see also Figure 2.2, below). Also, note that when an interglacial reaches its peak, it descends *very quickly,* in geological time, into its cooling cycle. If we must fear climate, then we should fear the "tipping point" back into glacial conditions, not further warming.

The 100,000-year glaciation cycle

When it began 2.6 million years ago, our Ice Age operated on a 41,000-year cycle, coincident with the planet's 41,000 "tilt" cycle (for more details on these cycles, see **Sidebar 1: The Milankovitch Cycles** at the end of this chapter).

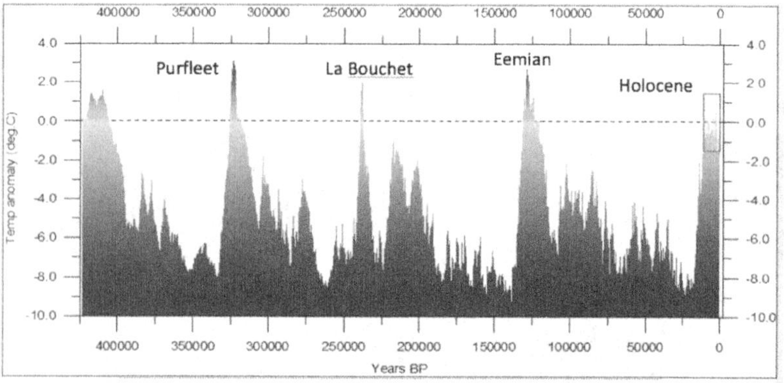

Figure 2.2: Vostok ice cores from Antarctica show that previous interglacials were warmer than our interglacial has reached so far, without human-caused CO2 emissions. In other words, our current warmth is well within 'natural variation'

For the past million years or so, though, our Ice Age has taken on a 100,000-year cycle: roughly 80,000-90,000 years of glaciations followed by 10,000-20,000 years of interglacial warm spells (see Figure

2.2). We are currently in one of the relatively short-lived interglacials, the Holocene, in which the Earth's climate is livable in most regions.

This fact is worth emphasizing, by the way. Over the past million years, in particular, our planet has been bitterly cold for *80-90 per cent of each 100,000-year cycle*; the warm interglacials are relatively short: 10-20 per cent of the cycle. We are *lucky* to be in an interglacial; we will be very *unlucky* when the interglacial ends and encroaching glaciers grind our civilization to dust (if we haven't already destroyed it with Net Zero).

When glacials dominate (the jagged downward dips in the two lines on Gore's chart and in Figure 2.2), much of the Northern Hemisphere is covered by ice, fringed by thousands of square miles of tundra.

As part of the most recent glacial cycle, Earth was ice-bound a mere (in geological terms) 20,000 years ago, and broke into an interglacial a mere 12,000 years ago. It was this warming that allowed agriculture and civilizations to arise; the warming was, again, part of the planet's natural climate variations. It wasn't in any way caused by humans.

Fun fact: The three previous interglacials (and likely interglacials before that) were *warmer* than we've reached so far in our interglacial, the Holocene (Figure 2.2). The three previous interglacials also had *higher sea levels* than today. For example, in the most recent interglacial, the Eemian of 125,000 years ago, temperatures were 1-3°C warmer than today, with sea levels 4-6 metres (14 to 20 feet) *higher* than today's sea levels.[2]

Crucially, these Eemian warmer temperatures and higher sea levels occurred *without* CO_2 enhancement from industrial fossil-fuel burning, cities or agriculture. In short, our Holocene interglacial is still well within "natural variation" compared to earlier interglacials in terms of temperature and sea levels.

Indeed, if past interglacials are a guide, warming and sea level rise would occur today *whether we were emitting carbon or not*. This is worth thinking about when alarmists point to Holocene temperatures and sea levels as somehow aberrant.

Gore gets order of temperature and CO_2 wrong

Getting back to Gore's chart (Figure 2.1): The top squiggly gray line is carbon dioxide, the bottom white squiggly line is temperature. As you can see, the two lines fit each other pretty well, except up to recent times when CO2 takes off into the stratosphere (as it were).

The lesson Gore wants us to draw is that in the past 650,000 years, when CO2 went up, temperature also went up. When CO2 went down, temperature went down. Since the peaks and valleys in CO2 levels appear to more or less fit peaks and valleys of temperature, then clearly as CO2 levels changed, so did temperature in response. Strong correlation established! Alarmist "control knob" climate theory confirmed!

As Gore explains in the book version of *Inconvenient Truth:*

> It's a *complicated* relationship [between CO2 and temperature], but the most important part is this: When there is more CO2 in the atmosphere, the temperature increases because more heat from the Sun is trapped inside.[3] [emphasis added]

Seems simple enough. But there's a scientific problem here that Gore ignores (because it doesn't support his alarmist case).

If CO2 levels went up and down several times over 650,000 years (and, of course, earlier than that), leading to ups and downs of temperature as part of a roller-coaster series of glacials and interglacials, then the next logical question is: *why* did CO2 levels go up and down? What *caused* these changes in CO2 levels that, in turn, produced such radical changes in temperature?

In particular, 12,000 years ago, what caused CO2 levels to suddenly shoot up enough to warm the planet and end the most recent glaciation? It wasn't humans: neither agriculture nor industry nor cities had been invented yet.

Gore is cagey in his choice of words on this issue (the relationship of CO2 and temperature is "complicated") for a very good reason: the scientific *facts* show that Gore is wrong. Dead wrong.

Over the 650,000 years in his wallchart, ice-core records[4] show that changes in temperature came *first, followed by* changes in CO_2. In fact, over this time, changes in CO_2 levels *follow* temperature changes by *several hundred years* (800 years is a frequently cited figure).[5] Why? Because, as we saw in Chapter 1, when the planet warms, the oceans warm up and release carbon dioxide. When the planet cools, the oceans also cool and slowly re-absorb carbon dioxide.[6]

And note that, during our 2.6-million-year Ice Age, temperature and CO_2 have had *the same relationship* as the cooling over the past 50 million years that we saw in Chapter 1—*first* changes in temperature, *then* changes in CO_2 levels.

Gore's rather basic scientific error—he doesn't state outright that CO_2 changes *caused* temperature changes over 650,000 years, but this is clearly implied in his presentation—seems unbelievable.

The wallchart is the core image of the movie—how could Gore think such a bone-headed misreading of the facts would get by actual scientists? Especially if, as he claims, he had scientists advising him in making the movie? How did all these scientist-advisors allow the movie to go ahead with a central claim, even if strongly implied rather than stated outright, that is completely wrong, and known to be wrong by other climate scientists, including Gore supporters?

Even Gore supporters know he is wrong

For example, here is John Houghton, a prominent climate alarmist, Gore supporter, and a lead author for the IPCC, on Gore's error:

> That carbon dioxide content and temperature correlate so closely during the last ice age is not evidence of carbon dioxide driving the temperature *but rather the other way round.* ... I often show [Gore's] diagram in my lectures on climate change but always make the point that *it gives no proof of global warming due to increased carbon dioxide.*[7] [emphasis added]

Even Andrew Weaver, Canada's leading climate alarmist, admits that Gore is wrong on this point, writing:

> Yes, *temperature changes came before carbon-dioxide changes in the recent glacial record.* Temperature change was driven by small changes in the Earth's orbit [i.e., the Milankovitch cycles, discussed in **Sidebar 1: The Milankovitch Cycles**]."[8] [emphasis added]

The only part Gore might, maybe, possibly have gotten right is at the very end of the chart, when CO2 levels zoom, leading to his claim that these higher, mainly human-caused levels of CO2 could cause dangerous warming.

We'll explore that relationship (or non-relationship) more fully in Chapter 4, but for now, Gore is asking us to believe, as he stands on a lift to point to the zooming CO2 level (while temperature is zooming up but not nearly as much as CO2), the following:

For the past *650,000 years*—and, by extension, the past *50 million years* and more—temperature and CO2 behaved one way, with temperature influencing CO2 levels. But, then, in the last 200 years, *the laws of physics suddenly changed so that CO2 now determines temperature.*

In other words, Gore is proposing two universes, each with its own physical laws. On Gore's Wonderland side of the looking glass, CO2 changes cause warming and cooling in the 650,000-year geological record and, presumably, long before that. On the Real Science World side of the looking glass that most physicists and even some climate scientists believe in, warming and cooling caused the CO2 changes.

I don't think even Aristotle could "harmonize" this disconnect between the two worlds, although alarmist climate scientists seem have no problem when the facts don't fit with their theory—they go with the theory.

It's *possible* Gore is right, that the laws of physics have suddenly been suspended at the alarmists' behest, but unlikely. In Real World physics, there is no firm *scientific evidence* to support Gore's claim, and the IPCC's claim, that CO2 is now driving temperature to hitherto unseen heights.

Quite the contrary: on the far right side of Figure 2.1, meaning the present, the white temperature line is up but certainly *not* following the grey CO2 line into the stratosphere. And in fact (as we'll see in Chapter 4), in 2006, when Gore's movie was showing in theatres, the global temperature had basically flatlined.

Eight other errors in Gore's movie

If Gore has a major scientific error, or at least misdirection, in the main argument of his movie and accompanying book, one might reasonably ask: are there other errors? Turns out, there are.

In 2007, a British parent objected that *An Inconvenient Truth* was being taught in his child's classroom as sound, scientific fact, rather than political propaganda, and he sued. He won his case as a British judge found nine major scientific errors in the film. Just for the record, the judge found that:

1. Sea levels won't go up seven metres (23 feet) in the near future due to melting of the Arctic and Antarctic ice caps, as Gore says. Even the IPCC estimates that sea levels might only go up perhaps half a metre (1½ feet) in the next century, at worst.
2. The Pacific atoll islands aren't in danger of sinking under the ocean, and whole island populations have not been evacuated, as the film says.
3. Very few climate scientists believe that global warming will shut down the Gulf Stream and cause another ice age.
4. Gore's huge temperature-CO_2 wall graph has cause (temperature) and effect (CO_2 levels) reversed [we've already seen this].
5. Global warming is not threatening polar bears or causing them to drown. Only four have drowned that we know of (at the time of the judge's ruling), but that was due to a storm, not global warming, and all but two of 17 world polar bear populations were either stable or increasing at the time of the judgment.[9]
6. The glaciers on Mt. Kilimanjaro in Tanganyika aren't melting because of global warming. The Kilimanjaro glacier has actually been melting since the 1800s, so industrial carbon emissions can't be the reason, and most

scientists think surrounding deforestation is causing the problem.

7. The drying up of Lake Chad was likely due to reasons other than global warming.
8. The world's coral reefs are not bleaching due to global warming.
9. Global warming is not the reason for major weather events like Hurricane Katrina.[10]

In other words, virtually the entire movie is scientifically wrong, misleading, and/or exaggerated. Therefore, the judge ruled the film could be shown in schools, since there was some science in it, but had to be identified as having a strong *political* bias.

Alarmist scientists not embarrassed by Gore gaffes

If I was a climate scientist—or any scientist—I'd be embarrassed at having Gore's film associated with my profession. Yet this is not the case. Scientists greeted Gore's film with praise because he is getting out the right "message"—carbon dioxide is Bad. Wikipedia has comments from scientists who had seen the movie. A small selection:

- "[Gore] got all the important material and got it right."—William H. Schlesinger, dean of the Nicholas School of Environment and Earth Sciences at Duke University.
- "I sat there and I'm amazed at how thorough and accurate [the movie was]. After the presentation I said, 'Al, I'm absolutely blown away. There's a lot of details you could get wrong.'... I could find no error."—Robert Corell, chairman of the Arctic Climate Impact Assessment.[11]

"No error." "Got it right." From scientists. Wow.

For his part, Andrew Weaver gushed that he was "fortunate to attend" a talk by Gore in Victoria in 2007, and that Gore's speech was "passionate, inspirational and empowering."[12] If *Inconvenient Truth* had been shown as an individual presentation in a university climate class, with this many errors, I am certain the professor would have given it failing marks for lack of scientific accuracy.

A few scientists did notice a few errors, e.g., climate scientist Eric Steig, who observed that the film got the fundamental science right and the minor factual errors did not undermine the main message of the film.

A few "minor" factual errors? Nine? Including Gore's main exhibit (the wall chart), which implicitly reverses the scientific facts? And shouldn't a scientist care that *An Inconvenient Truth*'s message is *primarily* political, as the British judge found? An odd sort of "scientific objectivity," this alarmist climate science.

Alarmist climate scientists are very forgiving of their alarmist colleagues' errors, but not so forgiving of colleagues who aren't alarmist, as we'll see in Chapter 9 when we discuss Climategate.

'Imperfections' and 'technical flaws'

A major figure in climate alarmism, James E. Hansen, then director of the Goddard Institute for Spaces Studies (GISS), acknowledged that Gore's approach may have "imperfections" and "technical flaws," but "he has the bottom line right."

One of Gore's "technical flaws," for Hansen, was blaming increased storm intensity and frequency on global warming. Hansen concluded: "We [scientists] need to be more careful in describing the hurricane story than [Gore] is."[13] Good to know! However, Hansen doesn't enlighten us on any of the other "imperfections" and "technical flaws."

Apparently, for Gore's alarmist fans, getting the 650,000-year cause-and-effect of temperature and CO2 backward (although the order is implied rather than stated outright) is a "minor" factual error or "technical flaw," not to mention the other eight errors that the British judge noticed.

But, seriously: if Gore's case is so strong, if the "science" is so solid, why are there so many, indeed, *any* factual errors at all? Again, from Jean-François Revel: "How is it possible for a theory, which is false in its component parts, to be true as a whole?"[14]

There is a consistent pattern here: alarmist climate scientists are happy to ignore "flaws" in Gore's, and their own, science as long as they are getting out the right "message." That is, it's okay to torture the facts to make them agree with the theory, which is the opposite of good science. We'll further discuss the curiously flexible integrity of alarmist climate scientists in Chapter 9.

In the next chapter, we'll look more closely at the relationship, or lack of relationship, between temperature and CO2 over very long stretches of time—billions, not millions, of years.

Notes for Chapter 2

---

[1] National Oceanographic and Atmospheric Administration (NOAA), "What's the coldest the Earth's ever been?" *Climate.gov*, Feb. 18, 2021.

[2] IPCC 2007, Summary for Policymakers, "A Paleoclimatic Perspective," p. 9. The text reads: "The last time the polar regions were significantly warmer than present for an extended period (about 125,000 years ago), reductions in polar ice volume led to 4 to 6 metres (13 to 20 feet) of sea-level rise."

[3] Al Gore, *An Inconvenient Truth: The Planetary Emergency of Global Warming and What We Can Do About It*. Emmaus, Penn., Rodale Press, 2006, p. 67.

[4] Scientists use ice-core samples from places like Greenland and Antarctica to estimate temperatures and CO2 levels in the past few hundred thousand years. The levels of oxygen isotopes offer a rough measure of climate conditions at various times.

[5] See, for example, among many possible others, Hubertus Fischer, et al., "Ice Core Records of Atmospheric CO2 Around the Last Three Glacial Terminations." *Science*, March 12, 1999 (Vol. 283, No. 5408), pp. 1712-1714. Also, Nicolas Caillon et al., "Timing of Atmospheric CO2 and Antarctic Temperature Changes Across Termination III." *Science*, March 14, 2003 (Vol. 299, No. 5613), pp. 1728-1731.

[6] Edward Aguado and James E. Burt, *Understanding Weather & Climate*. Upper Saddle River, NJ: Pearson Education, 2004, p. 495. This is a university-level climate textbook.

[7] John Houghton, "The Great Global Warming Swindle." John Ray Initiative, March 8, 2007.

[8] Andrew Weaver, "Climate change is no conspiracy." Letter to the Victoria *Times Colonist*, May 24, 2007, p. A15.

[9] For accurate, non-alarming information about polar bear numbers, and other polar-bear related issues, see Susan J. Crockford's website *Polar Bear Science* at polarbearscience.com.

[10] The judge's full ruling can be found at http://www.bailii.org/ew/cases/EWHC/Admin/2007/2288.html. Wikipedia's discussion of the case can be found with the search terms "Dimmock v Secretary of State for Education and Skills."

[11] Wikipedia, "An Inconvenient Truth."

[12] Andrew Weaver, *Keeping Our Cool: Canada in a Warming World.* Toronto: Viking Canada, 2008, p. 110.

[13] William J. Broad, "From a rapt audience, a call to cool the hype." *New York Times,* March 13, 2007.

[14] Jean-François Revel, *Neither Marx nor Jesus.* New York: Laurel, 1971, p. 15.

Sidebar 1: The Milankovitch Cycles

Al Gore's *Inconvenient Truth* wants us to believe that the roller-coaster of glacials and interglacials over the past 650,000 years, and by extension the whole 2.6 million years of our Ice Age, was the product of changes in CO_2 levels (caused by what? He doesn't say). In fact, most scientists in the Non-Wonderland World believe the Milankovitch Cycles are to blame.

The cycles were proposed by Serbian geophysicist and astronomer Milutin Milankovitch in the 1920s. Based on years of calculations (by hand! No computer!) he suggested that the glacials and interglacials were caused by the constantly changing, cyclical and relative positions of the Earth and Sun, not by changes in CO_2 levels.

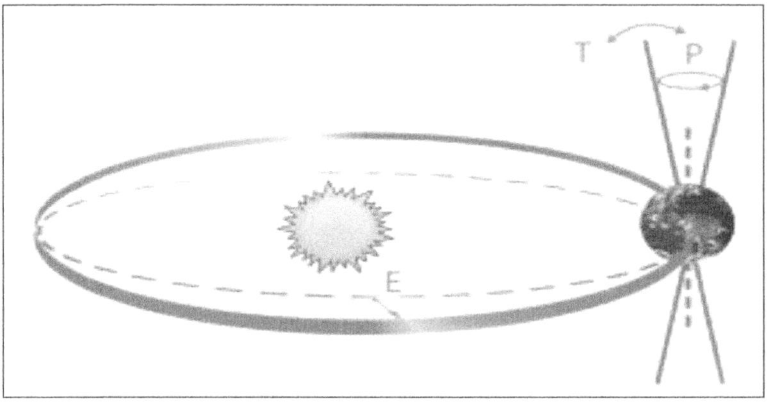

Figure SB1.1: The Milankovitch Cycles. E is the Earth's elliptical orbit, on a 100,000-year time scale; T is the planet's tilt, which varies over 41,000 years; P is the Earth's wobble, or precession, with a 23,000-year cycle. Source: IPCC

These cycles are based on three factors. One is the Earth's tilt, which varies between 22.1 degrees and 24.5 degrees on a 41,000-year cycle and causes the seasons. Secondly, the tilt wobbles like a top on a 23,000-year cycle, which is called the precession of the equinoxes.

Finally, there is Earth's slightly elliptical orbit around the sun, which has a 100,000-year cycle (see Figure SB1.1), sometimes more circular, sometimes more elliptical.

Before a million years ago, the cycle of glaciations/interglacials was 41,000 years, matching the Earth's "tilt" cycle. About a million years ago, the cycle evolved into 100,000 years, with 80,000-90,000 years of cooling/glacials, and 10,000-20,000 years of warming/interglacials (matching the Earth's 100,000-year elliptical orbit cycle). In other words, the Milankovitch cycles seem to underly the 100,000-year glacial and interglacial cycle.

Glacials begin when the three cosmic factors act together to produce cool summers in which, each year, less of the winter ice and snow melts. This can occur, for example, when the planet's elliptical orbit takes it farther from the sun in summer at a time when the Earth is also highly tilted.

Over thousands of years the ice builds up to a tipping point and a full-blown glaciation begins. On the other hand, unusually warm summers produced by a combination of the three factors begin to push back the ice, leading eventually to an interglacial.

But why did the Milankovitch Cycles, which presumably have been operating for tens of millions of years, take us into full glacier mode over the past 2.6 million years?

As noted in Chapter 1, the culprit wasn't carbon dioxide; it was tectonic changes, and most recently the closing of the Isthmus of Panama three million years ago. These changes gradually cooled the planet to today's Ice Age, *which is the coldest in 250 million years, and perhaps the coldest in Earth's 600-million-year history as a planet supporting life.*

Change in temperature first, followed by change in CO2 levels. Most scientists in the Real World agree on this, but alarmist climate scientists are happy to let the public believe the opposite.

Chapter 3: CO_2 and temperature—the long view

Oh dear, what nonsense I'm talking!

We've already seen that, in the extreme right of Gore's wall chart, the line for CO2 goes up considerably, while the temperature line is also up, but only half as much (Figure 3.1).

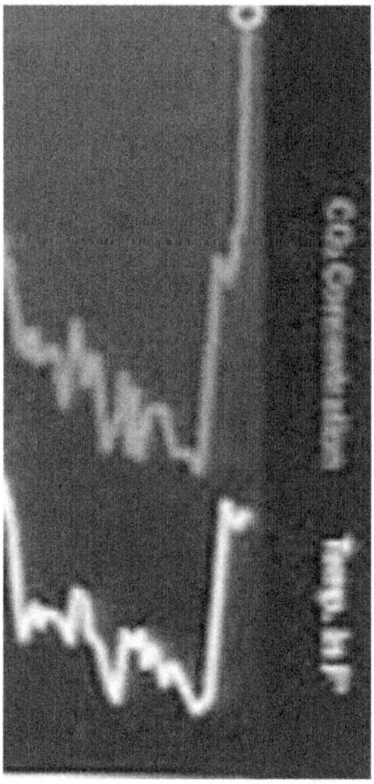

Figure 3.1: CO2 (top line) and temperature in present time in Gore's wall chart

Gore sees no problem with this: he apparently assumes that CO2 and temperature continue to be strongly correlated (thanks to their "complicated" relationship), and that the temperature will eventually catch up with the black dot at the top of the CO2 line.

But do note that, at least at the present time, the apparent tight correlation of CO2 and temperature that Gore wishes us to see no longer exists: CO2 is up and temperature is up, but temperature by not nearly as much.

There is another, more plausible interpretation than Gore's for this gap between temperature and CO2 in the present day, a time in which CO2 zooms upward at least in part due to human activities (industry, cities, farming and fossil fuels) while the temperature does not.

As we saw in Chapters 1 and 2, when temperature and CO2 appear to be correlated in the Real World (as opposed to Gore's Wonderland World), temperature change *precedes* changes in CO2 levels, not vice-versa; why would the CO2 peak in Gore's chart act any differently? Also, there are also several times in Earth's history when CO2 and temperature appear to have *no correlation at all.*

CO_2 and temperature over 600 million years

As evidence, have a look at Figure 3.2.[1] It's one of the most famous graphs in climate science and shows the relationship of temperature and climate over, not just hundreds of thousands of years, but 600 million years.

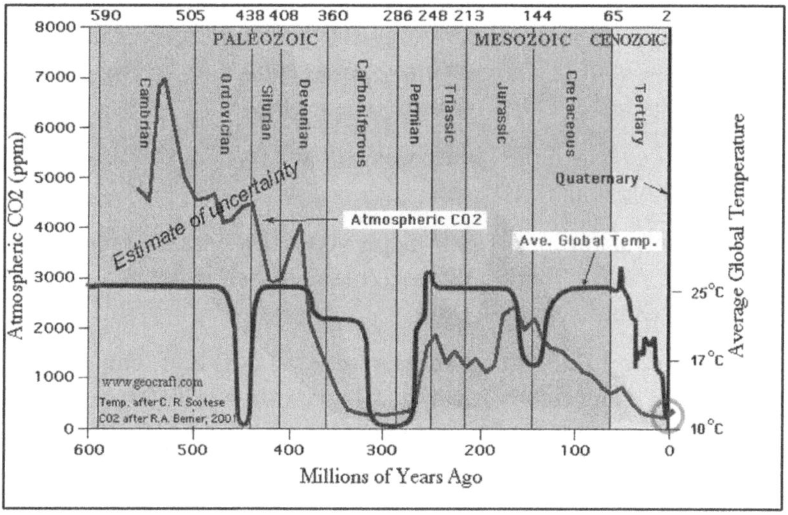

Figure 3.2: Carbon dioxide and temperature over 600 million years. Source: Scotese and Bernier. See chapter endnote for detailed reference

What do you notice about the relationship of temperature (lower thick line) and carbon dioxide (top, thinner line)?

Well, for a start, they're not very well correlated at all. There are times, such as 450 million years ago, when CO2 was very high (so the planet should have been super warm!) but the temperature was very low; in fact, at that time Earth was in one of its worst Ice Ages. This lack of correlation rather strongly implies that a variety of natural (as opposed to human-caused) factors other than CO2 are the "control knob" of climate, assuming there is a "control knob."

Happily for the alarmist theory, CO2 and temperature were both low 275 million years ago (the Ice Age before ours), so apparently correlated. But, alas, after that temperatures were quite high while CO2 levels fluctuated and eventually fell. This is especially true in the Eocene, 55 million years ago, when the planet was as warm it's been in recent geological history (the PETM noted in Chapter 1) but CO2 was falling.

In fact, 150 million years ago CO2 started falling while temperatures remained high for tens of millions of years. If there is, as Gore and alarmist climate science claim, a strong (if "complicated") correlation between CO2 levels and temperature—if CO2 is the "control knob" of temperature—*shouldn't temperatures have fallen in lock-step with the start of the fall of CO2 150 million years ago*?

Or, alternatively, since the planet was still very warm, shouldn't CO2 levels have stayed high?

The CO2 "control knob" theory is disproven right there—it's clear that other climate factors than CO2 are to blame for changes in temperature (a.k.a., "climate change").

In the bottom right of Figure 3.2, in our time, CO2 and temperature do once again come tightly together (the last time was 250-300 million years ago—the previous ice age).

Current ice age coldest in 500 million years

The bottom right of Figure 3.2 represents the glacial roller-coaster of the past 2.6 million years, discussed in Chapter 2, when CO2 and temperature do seem to be in a relationship (although, to repeat—sorry, Al—temperature change *precedes* CO2-level change by as much as 800 years).

But keep in mind: as noted earlier, our planet is in the most severe Ice Age in 250 million years and perhaps longer! Even the National Oceanic and Atmospheric Administration (NOAA) acknowledges: "Although it has some competition from cold conditions occurring between 300 and 250 million years ago, *the most significant ice age in the last half a billion years may be the most recent.*"[2] [emphasis added]

Our current Ice Age could be the planet's coldest in 500 million years! That's definitely something to think about, especially for alarmists fretting about a degree or two of "global warming."

That changes of CO2 are the *ultimate source of temperature change* (the "control knob") is the foundation stone of alarmist climate science—the whole theory of dangerous anthropogenic warming falls apart if this isn't true, as does the theory that *reducing* CO2 levels will measurably reduce temperatures. Therefore, to keep this theory going, it's surely critical that temperature follow—closely!—increases or decreases in CO2 levels. The alarmist climate models certainly work from this assumption.

But as you can see from Figure 3.2, this is not the case over deep geological time. Nor, as we'll see in Chapter 4, is temperature dutifully following increased CO2 levels in present time at the extreme right of Gore's chart (Figure 3.1). No Aristotelian "harmonization" here.

So far we've been looking at periods of millions of years and hundreds of thousands of years. But if we zoom in for a closer look to our own time, as we'll do in Chapter 4, we'll see that, yes, the planet is getting warmer, but this warming isn't caused by carbon dioxide but by natural forces—the atmospheric and ocean currents acting to produce El Niño events.

Notes for Chapter 3

---

[1] Temperature reconstruction by C.R. Scotese; CO2 reconstruction after R.A. Berner & Z. Kothavala, "Geocarb III: A revised model of atmospheric CO2 over Phanerozoic time." *American Journal of Science*, Vol. 301, Feb. 2001, pp. 182-204.

[2] National Oceanographic and Atmospheric Administration (NOAA), "What's the coldest the Earth's ever been?" *Climate.gov*, Feb. 18, 2021.

Chapter 4: CO_2 is rising steadily: Why isn't warming?

It takes all the running you can do just to keep in the same place.

Carbon dioxide has been rising steadily over the past 1½ centuries as humans have industrialized the planet. This CO2 rise is shown in the "Keeling Curve," which scientists at Mauna Loa in Hawaii have been tending since 1958 (Figure 4.1). The curve is named for its creator Charles Keeling.

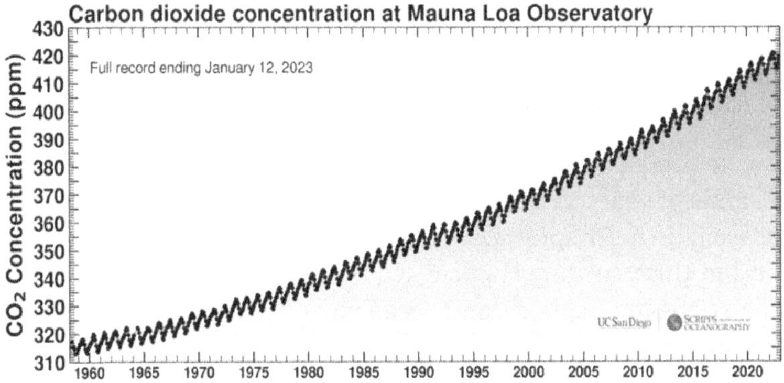

Figure 4.1: The Keeling curve showing steadily increasing CO2. Source: Scripps Oceanography

As presented in Figure 4.1, the Keeling Curve is basically a steady, roughly 20-degree upward progression.

If CO2 *precedes* temperature increases (i.e., is the "control knob" of climate), then we'd expect temperatures to follow the same pattern from 1958 to today: steadily rising temperatures on about a 20-degree or higher angle over the years.

And, no surprise, this is *exactly* what alarmist climate science expects! Most of the scientists' climate models show a world steadily warming at an angle of 20 degrees or more until we're all cooked (unless, of course, we stop our fossil-fuel-spewing ways). In other words, the climate models expect temperature to be highly "sensitive" and therefore highly correlated to increases in CO2 levels.

In Figure 4.2, below, the thick top straight line is the average of more than 40 climate-model warming estimates up to 2020; the lower straight line is the average of the *actual measured temperatures* back to 1975 (comparing the two measurements is called "hindcasting").

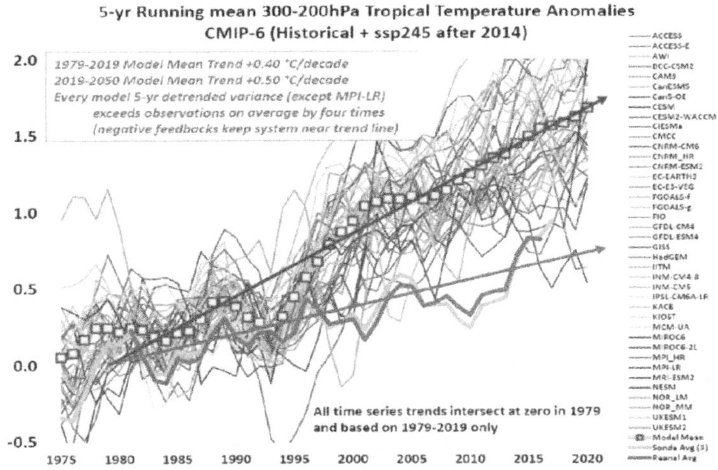

Figure 4.2: Comparison of models (top thick line is average) with actual temperature measurements (bottom straight line) by Dr. John Christy. Source: Clintel. https://clintel.org/new-presentation-by-john-christy-models-for-ar6-still-fail-to-reproduce-trends-in-tropical-troposphere

In other words, and alas for alarmist climate science, this present-day surge of temperature has not happened nearly as robustly as the models predicted. In fact, as we'll see in more detail below, for the past two decades the global temperature has remained fairly stable, indeed, almost flatline since the huge El Niño of 1998, with a few bumps upward, and then flat again.

Granted there may be, in Al Gore's Wonderland World, a lag—years? decades? centuries?—between CO2 increases and temperature, just as, in a different, Real World universe with different laws of physics, temperature changes caused changes in CO2 levels over 650,000 years, with a lag of about 800 years, as seen in Chapter 2.

Oh, wait: it's the same universe! Yet somehow the climate alarmist universe has different laws of physics than the realist universe. How can this be?

Models show much more warming than actual temperatures

Let's go back to Figure 4.2 to consider the grim future that the climate models predict for us.

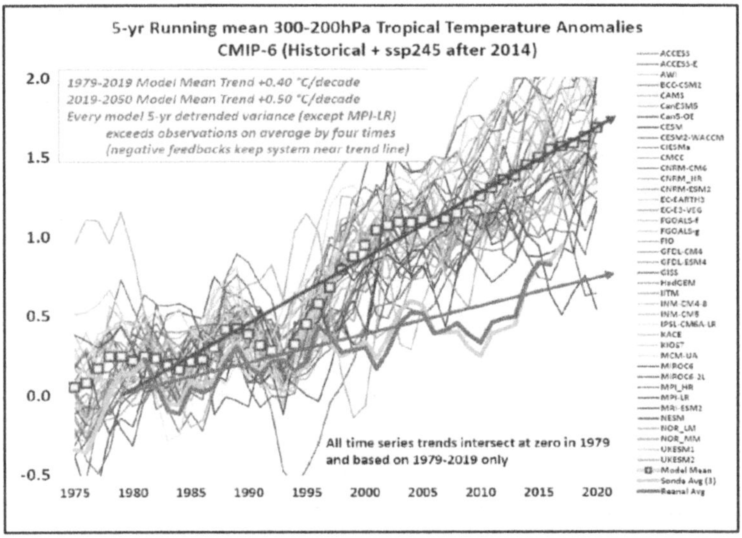

Figure 4.2: Comparison of models (average is top thick line) with actual temperature measurements (bottom lighter line)

As noted above, the rising top arrow is the average of temperature predictions by more than 40 climate computer models (the "spaghetti" lines; each strand represents a computer model).

Note that the angle of the average top line in the models is about 30 degrees, 10 degrees higher than the roughly 20-degree trajectory of the Keeling curve. That's because the models show what alarmist climate scientists *believe* will happen to temperature with CO_2 as the "control knob" of temperature. For alarmists temperature is very "sensitive" to CO_2 increases, so the angle of temperature increase is even *higher* than the CO_2 increase.

Despite the "certainty" of the alarmist climate scientists, however, temperatures in the first two decades of the 21st century (and earlier, as we'll see) have stubbornly refused to co-operate. As Figure 4.2 shows, temperatures have been below, and sometimes *waaay* below, the predictions of the computer models—the average actual temperature increase is an angle of about 12 degrees.

Using Woodfortrees for temperature data

Fortunately, you don't have to take my word for this. Believe it or not, you can become a climate scientist yourself! Or at least you can check the climate temperature data for yourself.

The website you need to visit is woodfortrees.org. **Sidebar 2: Woodfortrees** has details on how you can use this website to get temperature data going as far back as the mid-1800s.

Woodfortrees.org gathers the temperature data from five major climate-monitoring centres: The U.S.-based Goddard Institute for Space Studies (GISS); Britain's Hadley Institute (HadCrut); Remote Sensing Systems (RSS) and University of Alabama at Huntsville (UAH), both U.S.-based. (The fifth agency, Berkeley Earth Surface Temperatures—BEST—adds texture but not much light to the picture so we'll ignore it.)

The readings by GISS and HadCrut are land-and-sea-based; RSS and UAH use satellites. GISS and Hadley offer land-and-sea readings going back more than a century. Satellite climate data has only been available since 1979 (weather balloons were used before satellites).

That said, the satellites are generally considered more reliable than land/sea monitoring because satellite readings are less affected by external, non-climate factors ("noise").

For example, temperature-monitoring stations in or near cities can be affected by urban heating (e.g., concrete and asphalt) that can bias the readings toward "warming" but do not reflect natural "climate change."[1]

Temperatures from 2000-2023

With all that said, and using Woodfortrees.org to get the temperature data, let's have a look at the temperatures for the last 20 years or so, from 2000 to 2023—years of maximum climate alarmism—as measured by the four major agencies.

Figure 4.3 shows what the global temperature has been doing between the year 2000 and 2023. (Note that the charts show temperature *anomalies*—variations—from a previous 30-year baseline, not the actual temperatures.)

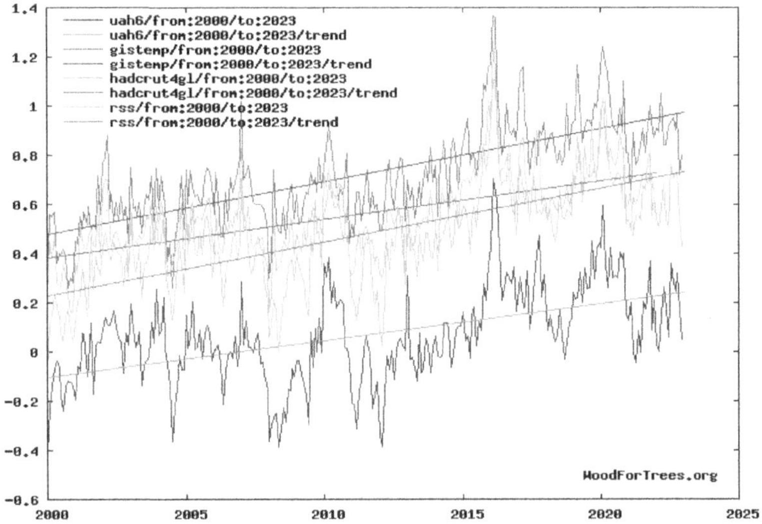

Figure 4.3: Temperatures 2000-2023 as reported by GISS, Hadcrut, RSS and UAH: Source: Woodfortrees.org

As you can see, the straight "trend" lines show a temperature increase of about 0.3°C in this 22-year period, or about 0.15°C per decade. If warming continued at this rate, in a hundred years the temperature might go up—wait for it!—1.5 degrees Celsius! Not 3 or 4 or 6 degrees, as the IPCC computer models predict, but 1.5 degrees. We can handle this.

No warming from 2000-2014

But let's break this time period down a bit more. How much "warming" was there between, say, 2000-2014? Oddly (given the "certain" nature of climate science), the temperature *barely budged* between 2000 and 2014. See for yourself (Figure 4.4).

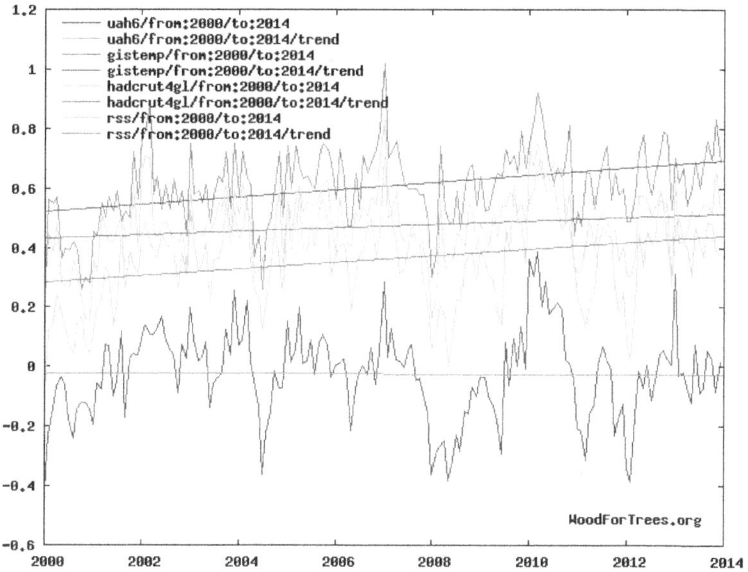

Figure 4.4: Temperature 2000-2014, all four agencies

Far from increasing considerably (even alarmingly!), as we've been led to believe by the climate doomsters, between 2000 and 2014 the temperature trend lines for three of the monitoring agencies rose by a fraction of a degree Celsius (basically flatline), and actually *fell* a bit in the UAH satellite readings (bottom line).

That wasn't supposed to happen! And warming of this level is certainly nothing to worry about even if the 2000-2023 trend was to continue for a century—again, perhaps a 1.5°C increase (or less) by 2120.

And, indeed, as early as 2005 (as we'll see in Chapter 7), it was becoming clear to climate alarmists that the 21st-century temperatures were refusing to obey the computer models, as they should if CO2 was the "control knob" of climate. This disconnect between theory and reality caused considerable consternation in the alarmist climate community (again, more details in Chapter 7)!

No warming from 2017-2023

Just for fun, let's take a look at another temperature period: the last six years or so (at the time of writing), 2017-2023, on all four agencies (Figure 4.5). Once again we get a surprising result: all four agencies show a flat or *falling* temperature trend. Again, this is emphatically *not* what we'd been led to expect by alarmist climate scientists, even in the short term.

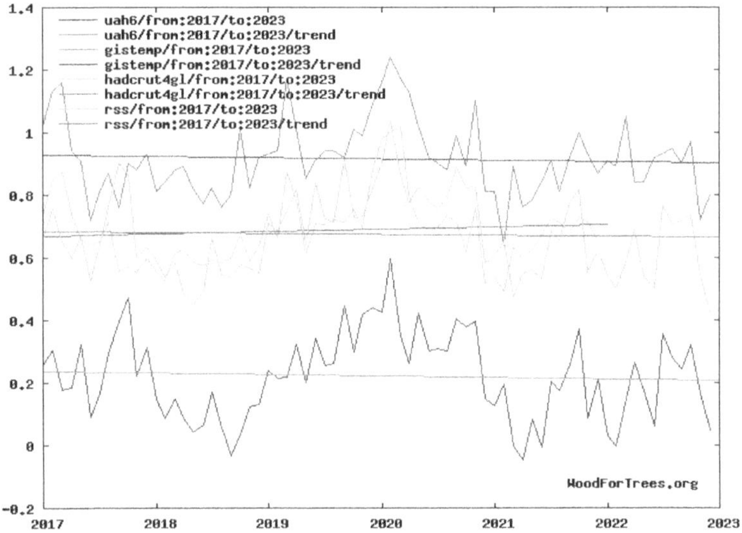

Figure 4.5: Temperatures 2017-2023, all four agencies

The U.S.'s National Oceanographic and Atmospheric Administration (NOAA) has a nifty gadget on its website that allows you to compare global land and sea temperatures today against a 1901-2000 anomaly baseline (the chapter endnote has the website address).[2]

As you can see from Figure 4.6, NOAA's data shows warming of +0.14°C/decade (rather than flat) when it comes to the 2000-2014 temperatures; this is because its anomaly baseline is the entire 20th century. The 20th century began a degree Celsius colder than today, so NOAA will show more anomalous warming than the late 20th century, 30-year baselines of the other climate agencies. Still, nothing to be alarmed about, you'd think.

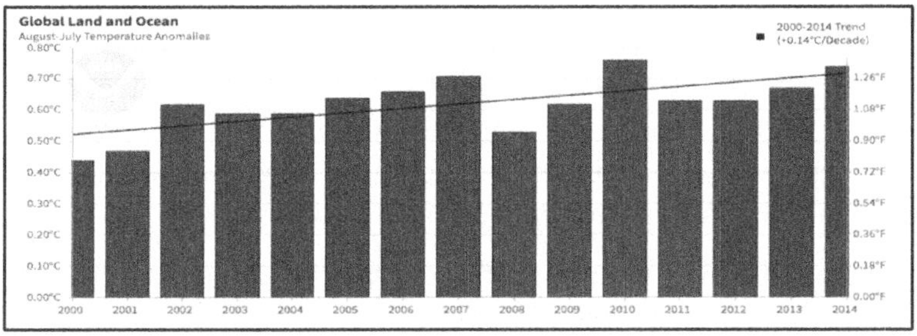

Figure 4.6: NOAA shows warming of +0.14°C/decade from 2000-2014. Source: NOAA Global Time Series

However, even NOAA shows 2017-2022 temperatures as flat-line, with a long-term August-July warming trend of—wait for it!—minus-0.01°C per decade, or basically zero degrees Celsius over a century (Figure 4.7, next page).

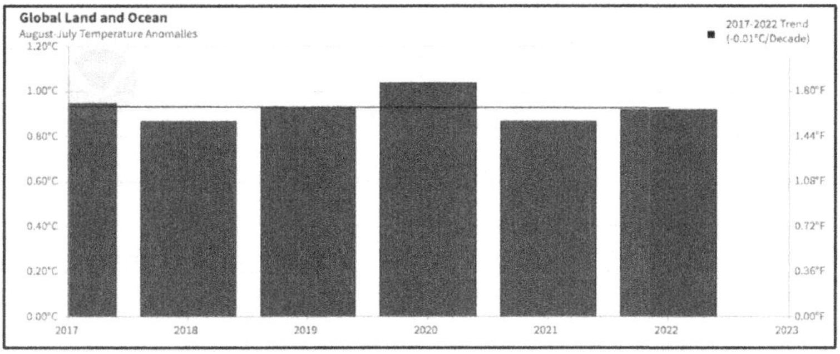

Figure 4.7: NOAA temperatures 2017-2022 (flat!)

Not daunted by reality, NOAA gamely continues to fill its website with stories aimed at creating climate alarm ("Fifth hottest January on record!", etc.).

You may be wondering why I skipped 2015-2016. There was a very strong El Niño in 2015-2016, which is why (the *only* reason why; CO2 need not apply) Figure 4.3 (2000-2023) shows an overall upward trend. However, this upward trend is a bit misleading.

It wasn't carbon dioxide but an El Niño spike of very high temperature (2015-16) that brought up the warming average for 2000 to 2022 (as shown in Figure 4.3).

Outside the El Niño years the temperature is pretty much flat, or even falling, for the two decades. By contrast, almost all of the computer models (Figure 4.2) predicted much more warming than has actually occurred in the temperatures outside the El Niño years.

Omitting the El Niño years also avoids charges of "cherry picking" the data to get the "cooling" result I might want, or the "warming" result the alarmists want. El Niños push the global temperature up. If a trend-line series begins at the height of an El Niño year, such as 2015, the trend line for the next few years will be "down," meaning cooling. This is also why the Figure 4.4 temperature trend line begins with 2000 rather than 1998 (also a strong El Niño year).

No warming from 1979-1994

Here's another temperature shocker you won't read in the alarmist climate textbooks but the facts are available using Woodfortrees: The planet barely warmed between 1979 and 1994. Yes, another 15-year period with virtually no warming while CO2 levels steadily increased.

Figure 4.8 (next page) shows the Woodfortrees temperatures for the four main climate-monitoring agencies (Hadley, GISS, RSS and UAH) from 1979, when satellite monitoring began, to 1994, three years before the large 1997-1998 El Niño (after 1994 the average starts to go up as the El Niño builds).

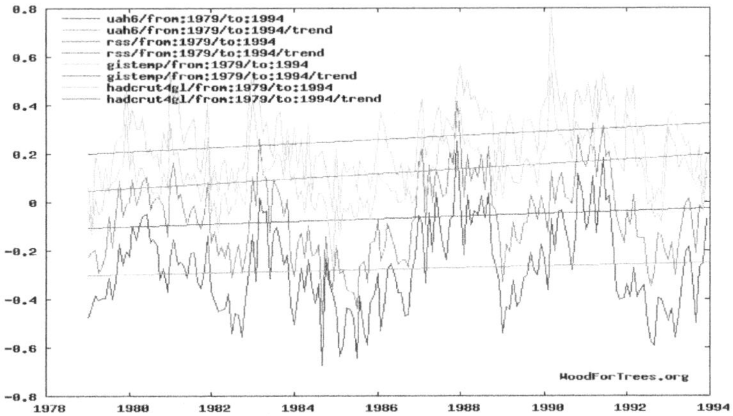

Figure 4.8: Four temperature monitoring agencies 1979-1994. Source: Woodfortrees.org

Once again, all four agencies are virtually flat. There's almost *no warming for 15 years* until the very large 1997-98 El Niño brings temperatures up.

Then "warming" goes flat again from 2000-2014, after which the El Niño of 2015-16 brings the temperature average up (to the everlasting relief of climate alarmists everywhere!), followed by flat to 2022/23 (the time of writing).

Figure 4.9, below, highlights how the 1998 El Niño, not carbon dioxide, bumped the average temperature up from 1979 to 1998, followed by a flat reading from 2000-2012 (and, really, up to 2014), at which point a new El Niño was building, creating the illusion of *CO2-caused* warming overall.

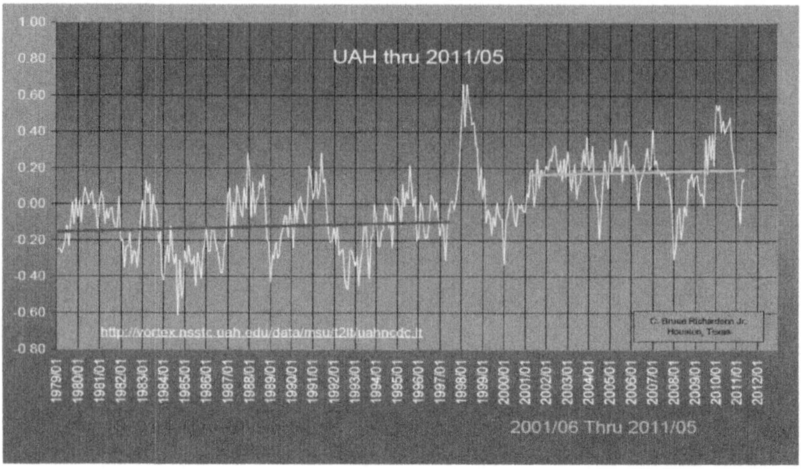

Figure 4.9: Average temperatures as shown by UAH from 1970-2012. Temperatures are flat until boosted by an EL Niño event. Source: C. Bruce Richardson

El Niños are weather, not climate

Now, climate-wise, El Niños are *weather*, not climate, events; the *climate* is what happens *between* the El Niños. And as you can see from Figures 4.4 to 4.9, the *climate*, from 1979 to 2023, is pretty much flatline: either no warming or very little.

This *isn't* what we'd expect if carbon dioxide was the "control knob" of temperature, especially in the last 50 years when CO2 levels have been steadily going up. The CO2 "control knob" is rising but, outside of El Niño years, the temperature isn't—except in 2023, which has experienced "record" temperatures. But, once again, as we'll see in Chapter 5, these "record" temperatures are due not to CO2 increases but mainly to a large undersea volcanic eruption.

This flat-lining over several decades rather strongly suggests that carbon dioxide is not producing appreciable warming, much less dangerous warming. Why? How can this be? We'll discuss the reason when we look at a phenomenon called "carbon-dioxide band saturation" in Chapter 6.

What if we look at years before 1979? There was a big El Niño in the mid-1940s, so we'll start the next graph at 1947 and go to 1979, which is 32 years. Figure 4.10 shows the GISS and HadCrut readings for 1947-1979 (satellite readings only began in 1979):

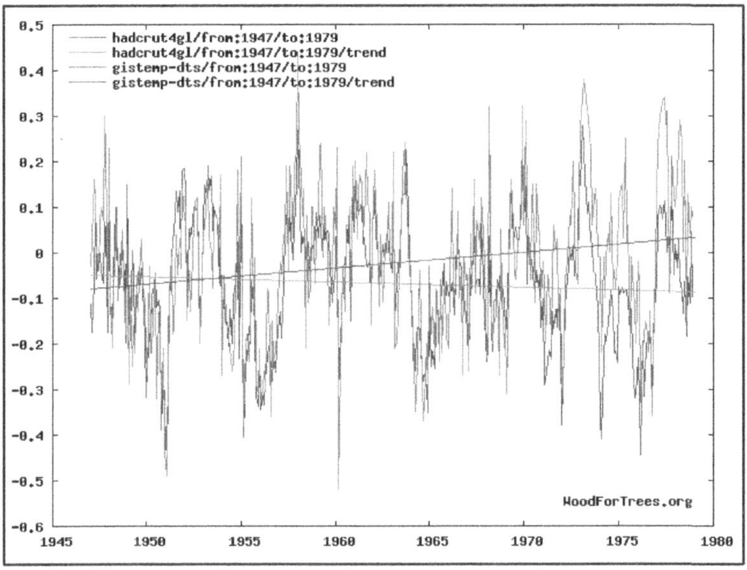

Figure 4.10: Temperatures 1947-1979. Source: Woodfortrees.org

The trend line for HadCrut (top line on the far left, lower line far right) shows temperatures slightly *falling*; the GISS trend line (top line at far right) shows temperatures very slightly rising, but basically flat. That's *thirty years* without appreciable warming.

One reason for this lack of warming is that the climate went cold from the 1940s to the late 1970s[3] (see Figure 4.11), to the point where some climate scientists (and a lot of the media) feared the start of a new Ice Age!

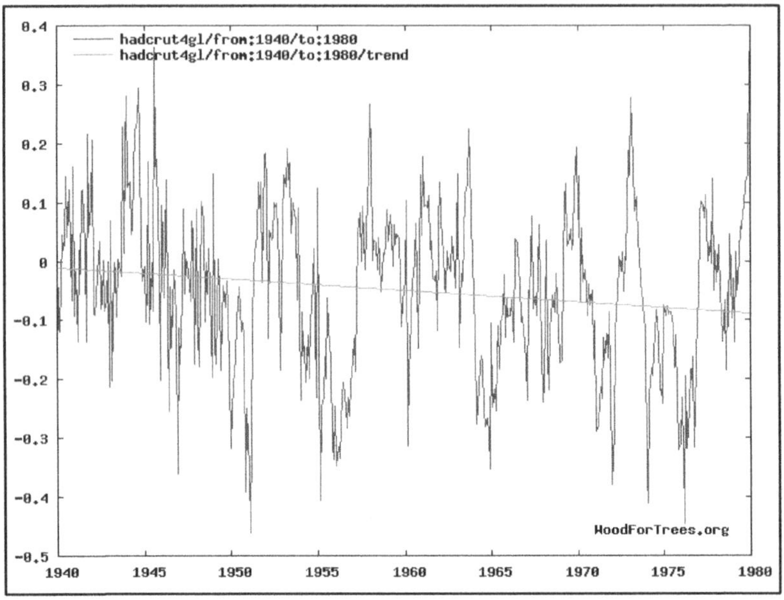

Figure 4.11: Temperatures 1940-1980. Going down!
Source: HadCrut global mean

For example, the late Stephen Schneider of Stanford University (a warming ultra-alarmist whom we'll meet again in Chapter 10) published a book in 1976 called *The Genesis Strategy* urging preventative action against a future ice age. He wrote:

> Today there are few people much concerned by the approach of the next ice age. And since ice ages take thousands of years to develop, why should we worry? There are several reasons to worry.[4]

So, in the mid-1970s Schneider was a leading alarmist about global *cooling*, and in the 2000s a leading alarmist about global *warming*. Honestly, you can't make this stuff up. That said, he was right about cooling, if that seems imminent. A return of the glaciers really would be "catastrophic" and worth getting alarmed about.

Summing up: We've been told for the past two decades that CO2 is causing "accelerated" warming and we're all (or many of us) going to die (the so-called "climate crisis"). The temperature data says otherwise.

The scientific explanation that makes the most sense to explain "global warming" (or lack of it) today—the explanation that is most "harmonious" with the facts—is that whatever has warmed the planet since the late 1700s, when the Little Ice Age ended (it was *really cold*; people could skate on the Thames River) and the globe began warming again, is also, perhaps, causing stronger than normal El Niños today. The question is, what caused the end of the Little Ice Age? Hint: It wasn't carbon dioxide.

Sunspots and temperature are correlated

Most scientists believe the depths of the Little Ice Age (about 1350-1850) were caused by a fall in solar radiation due to a decline in sunspots between about 1645 and 1715 (this time is called the Maunder Minimum)—fewer sunspots meant less heat from the Sun.

When the sunspots began to reappear, solar radiation increased and the Little Ice Age slowly began to warm up (Figure 4.12).[5] Note that the main causal factor in this "global warming" out of Little Ice Age was increased *solar radiation*, not increased carbon dioxide, although warmer temperatures will have coaxed more CO2 out of the oceans.

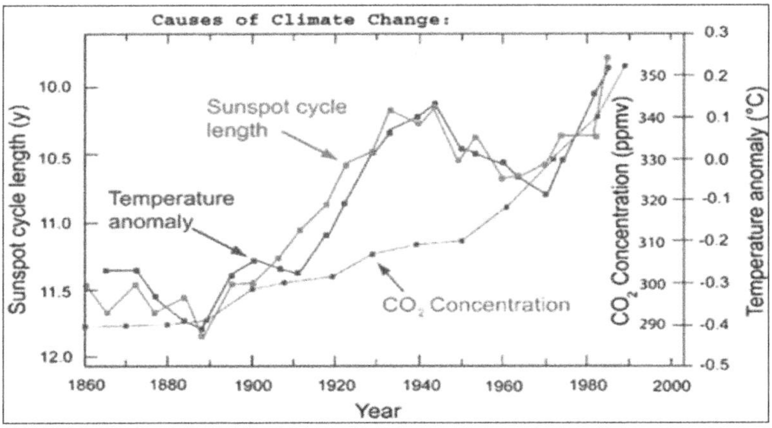

Figure 4.12: Sunspots and temperature from 1860-1990. Sunspots and temperature are more closely correlated than CO2 and temperature.

Just for the record, the IPCC does not believe solar variations are a significant factor in today's "global warming."

Back to El Niños: The primary cause of warmer El Niños isn't increased carbon dioxide, it's the cyclical interrelation of the Sun's radiation, air currents, and oscillating currents that warm or cool the oceans, which are the planetary heat sink. Humans may have a small role in making El Niños warmer, but through cities (the Urban Heat Island effect) and agriculture more than increases in CO_2 levels.

Climate blogger Bob Tisdale has done extensive research on the El Niño/La Niña phenomena, including ocean-circulation changes and how they influence global temperatures.

His conclusion? "It actually looks like the sea-surface temperature anomalies for the Atlantic, Indian, and West Pacific Oceans only warmed *during* ... El Niño events and then remained flat for the 9 to 11 years after them."[6] Translation: the Atlantic, Indian and West Pacific Oceans only warmed during El Niño years.

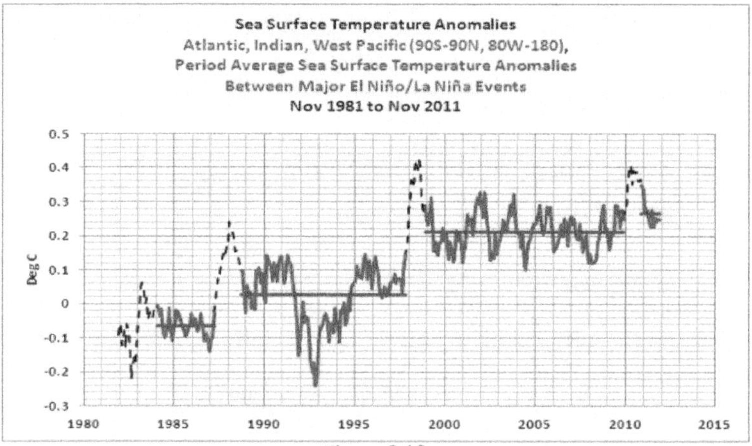

Figure 4.13: Warming more or less flat except after El Niño events. Source: Bob Tisdale

Figure 4.13, from Tisdale's research, shows how El Niños boost ocean temperatures (which in turn boost air temperatures) in stages every few years. Without El Niños the average temperature is pretty much flat, which more or less eliminates rising CO_2 as the primary suspect ("control knob") in warming. For a more detailed discussion of the El Niño cycles and how they are caused (hint: not by CO2 changes except perhaps very indirectly), see **Sidebar 3: El Niños** at the end of this chapter.

Tisdale also shows how El Niños and average temperatures have interacted from 1950 to the big El Niño in 2015.[7] Again, in Figure 4.14, we see non-warming between El Niño events, then "bumps" in temperature during El Niño years.

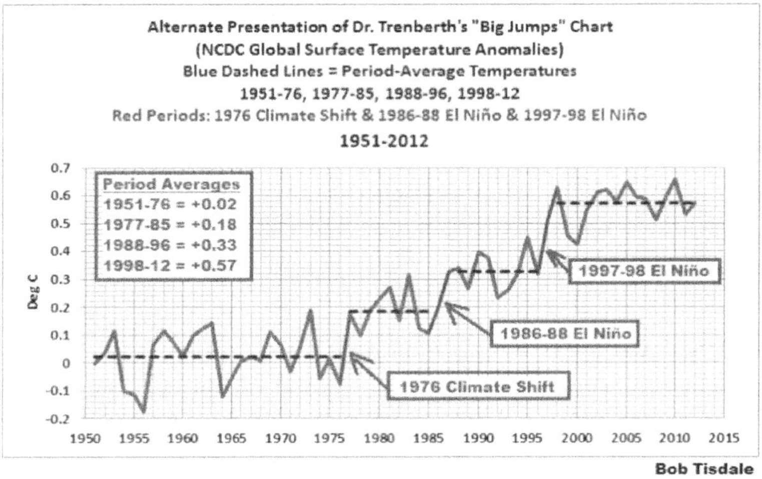

Figure 4.14: Temperatures flat until boosted by El Niño events. Source: Bob Tisdale

So, does increasing anthropogenic CO2 play a role in causing increasingly stronger El Niños? Possibly. During El Niños, according to NASA, more CO2 is released from the oceans, presumably adding to the overall warming effect.[8]

Or maybe not, with one science-news reporter noting:

> Based on reconstructions of Arctic climate variability in the greenhouse world of the Late Cretaceous, Southampton scientists have concluded that *man-made global warming probably would not greatly change the climatic influence associated with natural modes of inter-annual climate variability* such as the El Niño–Southern Oscillation (ENSO)

> or the Arctic Oscillation/ North Atlantic Oscillation (AO/ NAO)."[9] [emphasis added]

Translation: the various ocean temperature oscillations, including the El Niño cycle, don't seem to be related to human-caused warming (i.e., industry, cities and agriculture).

Between El Niños, temperatures are flat

What we see here is that, since at least 1940 and perhaps even since 1860 (referring back to Figure 4.12, the sunspot cycle and temperature), in the absence of an El Niño CO2's effect on global warming is, astonishingly, virtually nil, the *opposite* of what we're told by alarmist climate science.

The same seems to be true for cities and agriculture—these "anthropogenic" climate factors don't seem to be having much effect on temperature either. Again, in Chapter 6 we'll discuss carbon-dioxide "saturation," which explains why the warming influence of carbon dioxide appears to be on the wane.

To summarize: Once again, when we do a deep-dive into alarmist climate claims, we discover that the "global warming" that is supposed to be accelerating dangerously just simply isn't; outside of El Niño years the temperature is flat or only gently rising.

This means "The Science" (as expressed in computer models and the like) is not nearly as "certain" and "settled," nor as scary, as climate alarmists would like us to believe. And, in fact, the climate reality is the *opposite* of what climate alarmists would have us believe—the actual data simply does not show an "accelerated" or "catastrophic" potential outcome of "climate change."

In other words, increased CO2 isn't having much of an effect, if any, on current temperatures, which also means that *lowering* CO2 levels *will also have almost no effect on temperature*. In Aristotle's terms, we have another example here of a science that is not "harmonious" in its so-called facts.

But, again, you don't have to take my word for it that temperatures have flat-lined, or almost, in the 21st century and even in the last half of the 20th. First, you can use Woodortrees.org yourself. Second, as we'll see in Chapter 7, alarmist climate scientists *themselves* are saying that these flat-lined temperatures are poking holes in their theory—and they are mighty concerned about it.

But first, to complete the picture, in Chapter 5 let's look at the *main* greenhouse gas (hint: it isn't carbon dioxide).

Notes for Chapter 4

---

[1] By "noise" I mean this: GISS and Hadley employ temperature-monitoring stations scattered about their respective countries and from a network of stations around the world. These are supposed to give a "global" temperature.

The problem is that a monitoring station that once gave fairly accurate *natural* temperature data—it was sited, say, in a grassy field or forest—may gradually have been encroached on by a town or a city or an airport. If someone has built, for example, a concrete parking lot nearby, the sensors will pick up the extra heat from the concrete and the station will therefore report that this location is getting "warmer." The data might show "warmer," but the "warming" isn't caused by changes in the climate. The monitoring agencies say they correct for this bias, of course, but these "corrections" may simply be adding yet more bias, humans being fallible and all (especially alarmist climate scientists with a shaky theory to promote).

Ocean-based monitors suffer from similar problems. First, there aren't enough monitors to give a reliable picture of ocean temperatures (ocean is 71 per cent of the Earth's surface, so really big!), and sea temperatures vary with depth and currents. It's hard to get an accurate standard "global" reading from sea temperatures. For our purposes here, it's enough to know that the land-and-sea-based sensors are good as far as they go, but are generally less reliable and accurate than the satellite sensors.

For more detailed information on the fallibility of ground weather-station sensors, see the website *Watts Up With That*, then search the tag "weather station." *Watts Up With That* has done extensive investigations into the reliability of these ground stations and found 96 per cent of them didn't meet the NOAA standards for accuracy.

[2] This temperature gadget is available at: https://www.ncei.noaa.gov/access/monitoring/climate-at-a-glance/global/time-series.

[3] This three-decade cooling trend is usually attributed to air pollution, which has a cooling effect. The increase in warming in the 1980s is sometimes attributed to successful efforts to clean up air pollution. In other words, it wasn't carbon dioxide increases that caused warming, but anti-pollution measures that reduced cooling.

[4] Stephen Schneider, *The Genesis Strategy*, New York: Delta, 1976, p. 66.

[5] The graph comes from Eigil Friis-Christensen & K. Lassen, "Length of the solar cycle: An indicator of solar activity closely associated with climate." *Science*, Nov. 1, 1991, pp. 698-700. Carbon dioxide concentration was added by Friends of Science website.

[6] Bob Tisdale, *If the IPCC was Selling a Product, Would the FTC Stop Their Deceptive Ads?* 2012. Kindle e-book, location 1311.

[7] Tisdale, "The 2014/15 El Niño – Part 6 – What's All The Hubbub About?..." *Watts Up With That?,* April 30, 2014. See also Donald Rapp, "El Niños and La Niñas and Global Warming." *Climate, Etc.,* May 7, 2014. Both offer detailed discussions on the possible relation between carbon dioxide and El Niños; both conclude the connection is possible but not significant.

[8] NASA, "El Niño gives the world gas." November, 2007. The article notes: "Scientists notice significant changes in the air during an El Niño year. Concentrations of two important greenhouse gases, methane and carbon dioxide, go up significantly." Also, El Niños increase the chances of wildfires, although "climate change" usually gets the blame.

[9] Andrew Davies, et al., "Tropical ocean-atmosphere controls on inter-annual climate variability in the Cretaceous Arctic." *Geophysical Research Letters* 38. Also, found at *Watts Up With That*, "New paper: Unlikely that man-made global warming would cause a permanent El Niño state." Feb. 11, 2011. The other cycles—AO/NAO—are alternating warm/cold currents in the Atlantic and Arctic. El Niño cycles (ENSO) are primarily Pacific phenomena, although they do affect the weather across the planet.

Sidebar 2: Using Woodfortrees to find temperature data

The website woodfortrees.org offers a gadget that allows anyone to access global temperature changes (the ups and downs) going back to the mid-1800s.

Note that woodfortrees doesn't give the actual temperature; it is the temperature compared to an earlier baseline 30-year period, such as 1970-2000 or 1980-2010, hence "anomaly" or variation from. That said, the temperature anomaly does give you a good idea of how the global temperature is faring (up or down).

To learn what the temperature has *actually* been doing, as opposed to what climate alarmists have told us it's been doing, go to the website's main page, click "Interactive" in the upper horizontal menu and you will get Figure SB2.1:

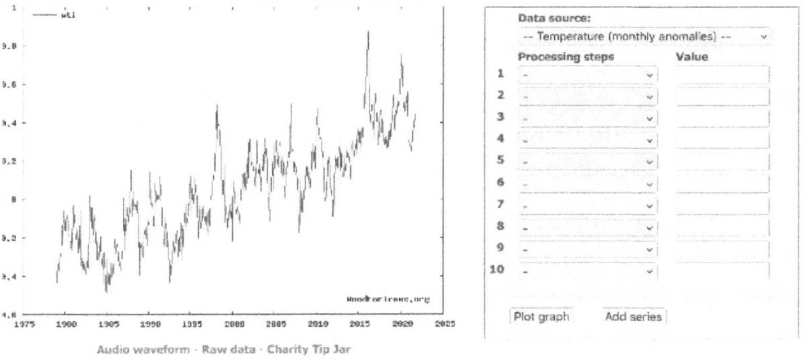

Figure SB2.1: Start page for 'interactive' menu

Clicking on "Temperature (monthly anomalies)" under "Data source:" will give you a list of the five major climate monitoring agencies and their various versions (see Figure SB2.2): Britain's Hadley

Institute (HadCrut); NASA's Goddard Institute for Space Studies (GISS, shown as GISTEMP); and Berkeley Earth Surface Temperature (BEST), all of which use temperature monitors on land and sea; and the two satellite-based monitoring services, Remote Sensing Systems (RSS) and University of Alabama at Huntsville (UAH). (I'm going to ignore the BEST readings because they are newer and pretty much duplicate GISS and HadCrut.)

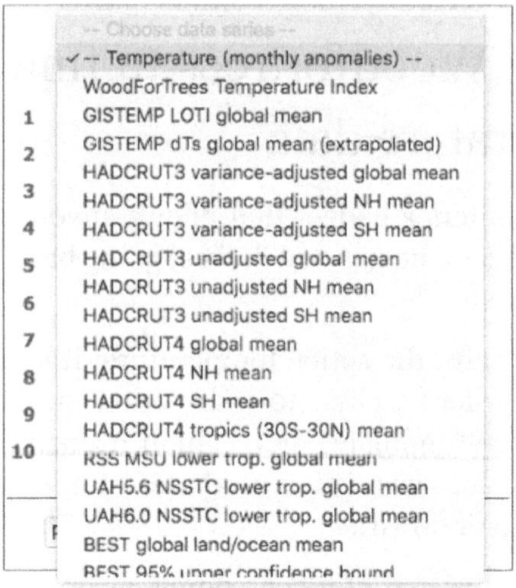

Figure SB2.2: List of climate monitoring agencies

Satellite-generated temperatures are considered more reliable than land-and-sea generated temperatures because satellites are less affected by local conditions (e.g., a land-based monitor near a concrete wall will show temperatures warmer than they actually are). However, satellite temperatures only go back to 1979.

Woodfortrees.org collects the temperature data from these five centres. It also offers an average of the five centres' data as its "WoodForTrees Temperature Index" (top of the list).

To get the temperature anomalies for any range of years, from now back to the mid-1800s for land-and-sea data, from 1979 for satellite data, you fill out the green box to the right of the temperature graph box.

Figure SB2.3: Getting data from one of the agencies

Click on a "Data source" (Figure SB2.3 uses the "UAH6.0 lower trop. global mean", a satellite reading). Under "Processing steps", click #1 "From (time)" and put in a date (in this case, "2000"). #2 is "To (time)", and I've put in "2021". Click on "Plot graph". That gives you a graph on the left side of the screen with a wavy red plot of the temperature anomalies.

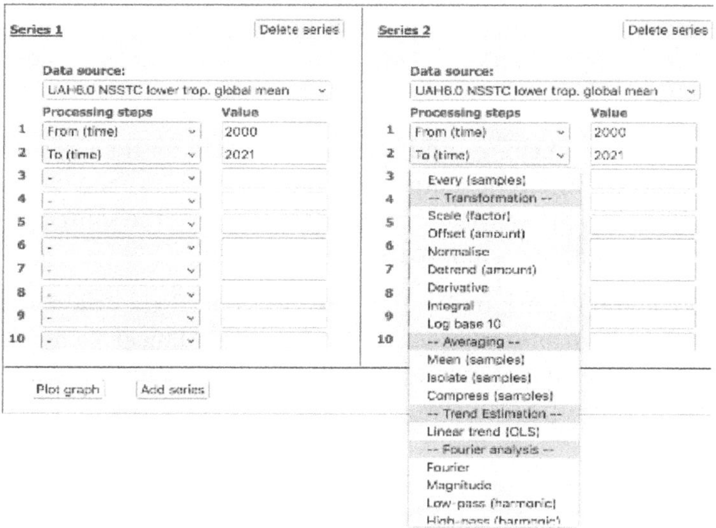

Figure SB2.4: Getting the trend line

Now what's needed is a trendline (Figure SB2.4, above). To get a trend, click on "Add series". You'll get another green box. Enter the same "Data source" and beginning (#1) and end (#2) dates, but in box

#3, a click will get you a list of options: choose "Linear trend (OLS)" under "Trend Estimation").

Click on "Plot graph" and you'll get the result shown in Figure SB2.5, with the temperature anomalies in red and a straight trend line in green. As you can see, the trend line is "up" for 2000-2021, thanks to the 2015-16 El Niño.

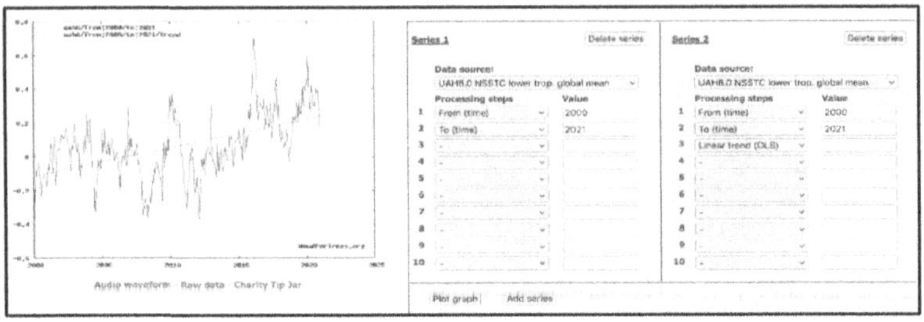

Figure SB2.5: Graph shows temperature changes plus a trend line

For comparison purposes you can display on one graph the temperatures and trend lines for two or more of the monitoring agencies' temperatures.

You do that by clicking "Add series" on the first of the green boxes, then choose a different "data source" and the "from" and "to" years. This time we'll choose "HadCrut 4 global mean" as the source (a land-sea reading) and the same years (2000-2021).

For a trendline, click on "Add series" and repeat the "data source," "from" and "to" information, then again pick "Linear trend" in #3. Continue for as many monitoring agencies as you want (yes, it is tedious). The set up for the two agencies, UAH and HadCrut, looks like Figure SB2.6:

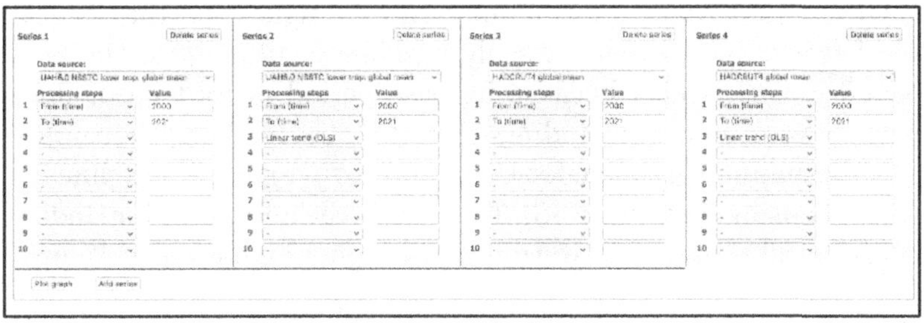

Figure SB2.6: Set up to show averages of more than one agency

If you include all four climate agencies, when you hit "Plot graph" you get the graph below, a comparison of the four temperature sets in different colors from 2000-2021 (Figure SB2.7). They pretty much agree on the ups and downs of temperature, while differing slightly on the vertical temperature scale due to their different anomaly baseline decades.

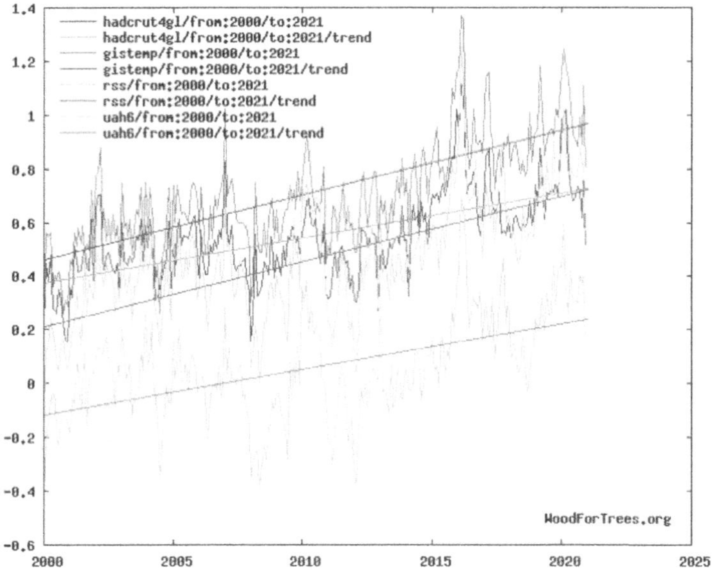

Figure SB2.7: Temperatures and trend lines for four agencies, 2000-2021

So, GISS's (top wiggly line) temperature is highest on the vertical scale because its 30-year anomaly baseline is 1951-1980. Hadley's (second down) is next because its base period is 1961-1990. Both baselines were colder (including fears of a new Ice Age!) than 1981-2010, the anomaly baseline for UAH, and 1979-1998, the baseline for Remote Sensing Systems (RSS). This means the anomalies for Hadley and GISS will show more "warming" (be higher on the vertical axis).

Figure SB2.7 shows an upward trendline of roughly 0.3°C over 20 years, or about .15°C per decade. Over a century that would be a temperature increase of 1.5°C, which is far from the "catastrophe" zone, unless you are a climate alarmist.

However, if you compare the four climate agencies for 2000-2014, just before the big El Niño of 2015-16, you will get the graph below (Figure SB2.8):

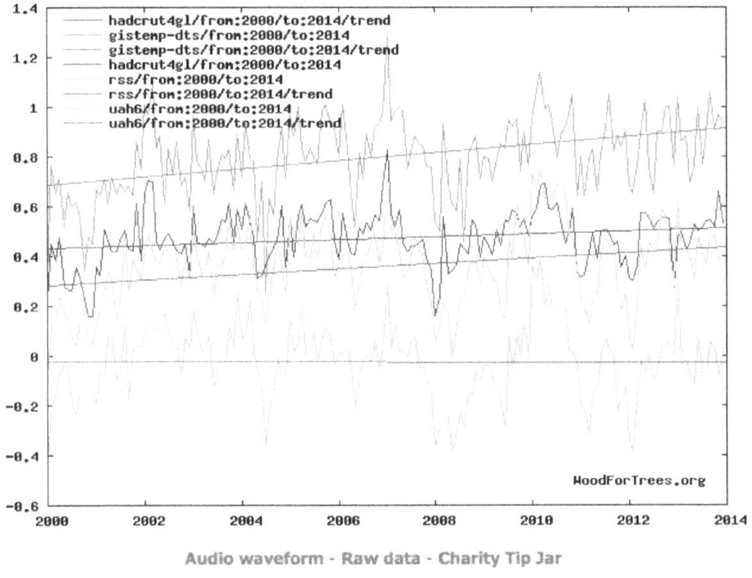

Figure SB2.8: Temperature anomalies for 2000-2014 for all four agencies: flat!

Looks pretty much flat, doesn't it? Yet during this decade and a half, the IPCC 2007 report described the evidence for global warming as "unequivocal" and, in 2008, IPCC chair Rajendra Pachauri told an

Australian audience that warming was occurring "at a much faster rate."[1] So in a decade and a half when the "global warming" frenzy reached new heights (Al Gore's *Inconvenient Truth* film came out in 2006), there was virtually no warming.

Don't believe me? Check for yourself on woodfortrees.org!

---

[1] Michael Duffy, "Truly inconvenient truths about climate change being ignored." *Sydney Morning Herald,* Nov. 8, 2008.

Sidebar 3: Are El Niños the 'control knob' of warming?

As we saw in Chapter 4, El Niños (not carbon dioxide) appear to be the "control knob" of climate, at least in the past century and certainly long before that. So how do El Niños work? How do we know they're not another product of higher CO_2 levels?

El Niños are not caused by or much affected by greenhouse gases/anthropogenic global warming. They've been going on for thousands of years, so human carbon emissions can't be the primary cause. The cause is the changing patterns of the Pacific trade winds.

El Niños are created when the trade winds in the Pacific Ocean blow more strongly than normal from east to west; this push builds up a reservoir of very warm water in the western Pacific (where the water is warm to begin with, again quite naturally).

When the trade winds ease, this reservoir of super-warmed water surges back toward North and South America, bringing warm and sometimes very warm temperatures. It's the strength of the trade winds, leading to greater or lesser concentrations of warm water in the western Pacific, not CO_2 (anthropogenic or otherwise), that determines the strength, and therefore the amount of warmth, of an El Niño.

The relationship of El Niños, ocean temperatures and trade winds is well (if complexly) explained in Bob Tisdale's book *If the IPCC was Selling Manmade Global Warming as a Product, Would the FTC Stop Their Deceptive Ads?*.

He uses the IPCC's own models, and other accepted scientific sources, to show that the oceans aren't that strongly affected by anthropogenic carbon dioxide—the oceans have not warmed nearly as much as the IPCC's models have predicted (this is true of land surface temperatures as well, of course).

The heat may well be “hiding” in the deep oceans, as some alarmists have claimed, but that is a theory, not an empirical fact.

Tisdale notes: "The actual sea-surface temperature data shows no influence from anthropogenic greenhouse gases. Since the mid-1980s, the sea-surface anomalies only rose during the El Niño events of 1986/87/88 and 1997/98—and likely as a result of the 2009/10 El Nino." (*If the IPPC was Selling* E-book, Chapter 3, section 5). Translation: human-caused carbon dioxide hasn’t had an influence on recent El Niños, but these El Niños have had a strong effect on temperature.

This means that when El Niños are strong, the planet's temperature rises—and then stabilizes at that plateau, or slightly less due to the subsequent, cooler La Niña. This has certainly been the case for the last 25 years since the very large 1998 El Niño—a big rise in warming at that time, and then a more or less flat line in temperature until the 2015-16 El Niño, then flat again.

If CO2 was as powerful as the climate alarmists claim, these flat-line “hiatuses” (as alarmists call them, with the clear message that they are temporary, move along, nothing to see here) between El Niños *would not occur*—warming would just carry on in a steady upward progression with the occasional small blip up or down. This is not the case.

Looking at temperature records for the last half of the 20th century, the same process seems to be at work (Chapter 4).

CO2-induced warming may have a small role in affecting both the strength of the trade winds and other climate phenomenon linked to El Niños, but CO2’s role is minor compared to the interaction of ocean and atmospheric currents that lead to El Niños in the first place.

Chapter 5: The 'extreme' global warming of 2023

"I wish I hadn't cried so much!" said Alice, as she swam about, trying to find her way out. "I shall be punished for it now, I suppose, by being drowned in my own tears! That will be a queer thing, to be sure! However, everything is queer to-day."

"'Something's changed': Summer 2023 is screaming climate change," reads a column headline in the Victoria *Times Colonist* on August 6. And, indeed, a screaming headline like this must warm the hearts of climate alarmists everywhere.

From all over the globe are reports of "record" temperatures, flooding, forest fires, and drought. The alarmists must be thinking: *Finally*! Finally the temperatures are obeying the models! That means those pesky "deniers" will have to admit it's time to dismantle Western civilization and move on to a "de-growth" society that abandons the evil fossil fuels that caused all this climate chaos.

Alas for the alarmists, what's primarily "changed" isn't human-caused increases in carbon dioxide but the eruption of a massive undersea volcano in January 2022 in the Pacific Ocean near Tonga that reportedly sent "a massive plume of water vapor into space." How much water vapor? 146 million metric tons.[1]

When I read this news story, I asked myself, "In the scheme of things, is 146 million metric tons a lot of water vapor? Or a drop in the bucket?" In other words, is 146 million metric tons of water vapor likely to affect the temperature significantly?

The news story didn't offer this important piece of information, but the answer appears to be yes, according to a paper in the journal *Nature Climate Change* and many other scientific sources.[2]

146 million metric tons increased the amount of water vapor in the atmosphere *by 10-15 per cent.* As a result it's possible, scientists say, that this one source alone could bump global temperatures above the dreaded 1.5°C increase from pre-industrial times, beyond which we'll all be turned into charred ash.

In other words, if we're looking for the source of the "extreme" temperatures of 2023, it's the Tonga volcano's massive addition to the water-vapor budget in the atmosphere, not the evil carbon dioxide. These volcanic additions to water vapor also underline a key theme of this book: there are many factors other than carbon dioxide, mostly natural rather than human, contributing to "global warming."

Also, in a kind of "perfect storm," 2023 is expected to be a big El Niño year, which means added warmth.[3] Fortunately, this two-pronged bump in "extreme" temperatures is expected to be temporary and global temperatures will return to "non-extreme" levels in a few years. So you've been warned: put your skeptic's hat on when you hear these claims of "record" warming caused by excess CO2 in 2023.

Water vapor is the main greenhouse gas

How could this addition of water vapor be so powerful? Because water vapor, not carbon dioxide (as we've all been encouraged to believe), is the *main* greenhouse gas, responsible for up to 90 per cent of global warming at any given time and place.

But what about carbon dioxide? We've been told (wrongly, but not surprisingly) that CO2 is the *main* greenhouse gas—it's the "control knob"! Surely CO2 is influencing 2023, too, isn't it?

Yes, CO2 is a greenhouse gas, one of several "trace gases" that help to warm the planet, including methane, nitrous oxide, and CFCs (chlorofluorocarbons like Freon, formerly used in refrigerators until banned internationally).

As "trace gases" they are found in tiny amounts in the atmosphere; CO2 is currently about 425 parts per million, or 0.043 per cent of the atmosphere—imagine four people and a new-born in a stadium that can seat 10,000.

However, the trace-gas warming power of CO_2 pales beside the warming power of water vapor, the stuff that makes rain and snow and fog.

To illustrate the warming power of water vapor: Moisture is scarce in the desert so the days are very hot, thanks to an unimportant (according to the IPCC) source of global warming, the Sun. But desert nights are cool or cold because there's no water vapor to "store" the Sun's heat.

By contrast, tropical regions are hot and humid day and night thanks to lots of water vapor ("it's not the heat, it's the humidity"). So the warming power of water vapor varies quite a bit depending on global location, but water-vapor-induced warming is considerable when atmospheric moisture present in large quantities (like after the Tonga eruption). More water vapor equals more warming.

Scientists tell us that our planet would be 33°C colder than it is without the greenhouse-gas effect. On an Earth that is 33°C colder than today's average of 15°C or so—around minus-18°C—all water would be frozen and the planet would be, basically, unlivable for humans and most complex beings (bacteria could survive). But how much of this 33°C is due to carbon dioxide, the all-powerful, Wizard of Oz "control knob"?

One commonly cited figure: if the world is 33°C warmer thanks to greenhouse gases, 30°C of that warmth is due to water vapor; the 3°C above 30°C is contributed by the trace greenhouse gases, and mainly CO_2.[4] In this reckoning, water vapor provides about 90 per cent of the atmosphere's warming potential, which seems about right to climate realists. If so, then CO_2's effect on climate, at only 3 per cent of total warming, is not great and certainly not the "control knob" of climate.

Alarmist sources prefer to downplay the warming potential of water vapor at more like 50 per cent,[5] others put the warming figure somewhat higher, but IPCC alarmists agree that the warming of recent decades is almost entirely due to CO_2, thanks to the "control knob" effect.

Here's how the CO_2-warming process works for the alarmists: as CO_2 increases, it traps more outgoing solar radiation, which creates additional atmospheric warmth (this is called "radiative forcing"; for more details on radiative forcing, and how the IPCC misuses this concept, see **Sidebar 6: Radiative forcing** following Chapter 13).

This added atmospheric warmth summons forth more water vapor, primarily through evaporation from the oceans. So as we humans add more CO2 to the atmosphere, we are also indirectly creating more water vapor, and hence more warming, in a vicious cycle, presumably to infinity.[6] (Fortunately, each increase of carbon dioxide has less and less warming power, as we'll see in Chapter 6.)

Why we don't try to reduce water vapor

That said, one might reasonably ask why the IPCC doesn't put more emphasis on water vapor as a greenhouse gas. Why aren't we, instead of trying to reduce fossil-fuel use (and risking ruin to our fossil-fuel-based economies), trying to reduce the amount of water vapor in the system? Wouldn't that bring warming down?

There are two basic reasons why reducing water vapor won't work as a cure for global warming.

One, humans can't control the levels of water vapor and therefore we can't control the amount of warming caused by water vapor. The amount of water vapor in the atmosphere depends on atmospheric conditions and, yes, even undersea volcanos, which is why we can't make it rain when and where we want it to, apart from the occasional successful rain dance.

And two, water vapor is a passive greenhouse gas; in itself it doesn't cause increased warming, but it can *amplify* warming from other causes (causes like CO2 but also presumably increases in solar energy, etc.). Without outside warming influences, and the positive-feedback warming loops they supposedly generate (alarmists tend to ignore the *negative*-feedback loops), the greenhouse effect from water vapor alone would be more or less constant.[7]

In other words, water vapor amplifies the effect of the other greenhouse gases like CO2 and methane that are lesser in volume but more active. The more water vapor, the more amplification. Hence, the "global boiling" (as the UN Secretary-General Antonio Guterres called it[8]) of 2023.

'Iris' effect of clouds regulates temperature

Fortunately, water vapor does more than heat up the atmosphere to dizzying heights of warming: as more water vapor gets into the atmosphere it also produces more cloud cover. Cloud cover (a.k.a., shade) *cools* the planet.

This natural regulator is called the "iris effect," so named because the eye's iris closes when there is a lot of light, opens when the light is dimmer. Similarly, more water vapor generates more warming, yes, but it also generates more cloud cover, which generates cooling. In this way, as it has done for hundreds of millions of years, the planet all by itself moderates the climate and avoids overheating (who knew?).

This "iris effect" is, of course, downplayed by climate alarmists, but may in fact be more powerful than the "greenhouse gas" effect, according to Dr. John Clauser, a co-winner in 2022 of the Nobel-prize for physics.

Clauser argues that there is no "climate crisis" in part because, referring to the "iris effect," the cooling action of clouds is "two orders of magnitude" more powerful than the warming power of CO2.[9]

Of course, climate alarmists continue to maintain that sinful humans and our uncontrolled carbon emissions will overwhelm the planet's internal ability to moderate the climate and put us all into a fiery grave. And this is, of course, an idea that should be taken seriously, if true.

Fortunately, once again, the alarmist view is exaggerated if not completely wrong. There is yet another force in addition to the "iris effect" that tends to keep warming due to CO2 in check: as CO2 levels increase, the warming power of each unit of CO2 diminishes exponentially, very quickly reaching near-zero. This is sometimes called the "CO2-saturation" or "carbon dioxide band-saturation" effect. We'll discuss CO2 saturation in Chapter 6.

Notes to Chapter 5

---

[1] Joel Achenbach, "Volcano eruption blew millions of tons of water into space." *National Post*, Dec. 13, 2022. See also *Phys.Org*, "Models show Tonga eruption increases chances of global temperature rising temporarily above 1.5°C." Jan. 25, 2023. There are many other articles online reporting this huge eruption and its effect on warming in 2023.

[2] Ayesha Tandon, "Tonga volcano eruption raises 'imminent' risk of temporary 1.5°C breach." *Carbon Brief*, Jan. 12, 2023.

[3] NOAA, "June 2023 ENSO update: El Niño is here." *Climate.gov,* June 8, 2023.

[4] Patrick J. Michaels & Robert C. Balling, Jr., *The Satanic Gases: Clearing the Air about Global Warming*. Washington, DC: Cato Institute, 2000, p. 25. See also David Archibald, "The Logarithmic Effect of Carbon Dioxide." *Watts Up With That,* March 8, 2010.

[5] Alan Buis, "Steamy Relationships: How Atmospheric Water Vapor Amplifies Earth's Greenhouse Effect." *NASA: Global Climate Change*, Feb. 8, 2022. Available at https://climate.nasa.gov/ask-nasa-climate/3143/steamy-relationships-how-atmospheric-water-vapor-amplifies-earths-greenhouse-effect/

[6] Buis, "Steamy Relationships." Cited above.

[7] Buis, "Steamy Relationships."

[8] UN News, "Hottest July ever signals 'era of global boiling has arrived' says UN chief." July 27, 2023. Guterres' words: "The era of global warming has ended" and "the era of global boiling has arrived."

[9] Andrea Wailzer, "Nobel Prize winner denounces alarmist climate predictions: 'I don't believe there is a climate crisis'." *LifeSite News*, July 13, 2023. Available online.

Chapter 6: The carbon-dioxide saturation effect

"That proves his guilt," said the Queen. "It proves nothing of the sort!" said Alice.

A basic law of physics is that as you put more carbon dioxide into the atmosphere, you get less warming effect from each additional molecule. How much less? The general consensus is about 50 per cent less warming per unit of carbon dioxide added; another way of saying this is that a doubling of CO_2 produces the same temperature increase as the original amount.

That is, if an additional 100 ppm of CO_2 gives you 1° Celsius of warming, the next 100 ppm will cause only 0.5°C of warming. The next 100 ppm will cause 0.25°C of warming. And so on. This is called the carbon-saturation effect or carbon-band saturation.

This process has been compared to a sponge. A dry sponge can absorb a lot of water; a wet sponge absorbs very little water because the sponge is "saturated."

Similarly, the atmospheric band that traps the extra warmth from additional CO_2 becomes "saturated" beyond a certain level of CO_2. There is additional warming out on the "wings" of the band, but this extra warming is quite small.

Here's a real-life example of how the carbon-saturation effect works.

The CO_2 level in the year 1900 was about 290 parts per million. At the end of the century, in the year 2000, the level was 370 ppm, for an increase of 80 ppm.

The IPCC reckons the global temperature increased by 0.6°C (+/- 0.2°C) during the 20th century,[1] with about half of that generally attributed to human activity, the other half due to natural causes (that said, listening to alarmists, you'd never guess that 95 per cent of the planet's annual CO_2 budget is produced *naturally* by the planet rather than by humans[2]). Other sources believe 100 per cent of the 20th-century warming was human-caused.[3]

So, just to be fair, let's assume that *all* of this 20th-century warming of 0.6°C was produced by human CO_2 emissions (it wasn't, but let's use the scenario that most favors the alarmists). So 80 ppm of additional CO_2 produced 0.6°C of warming.

But what would happen if we added another 80 ppm, raising the overall CO_2 level in the atmosphere to 450 ppm, a level we'll likely reach well before 2100. For climate alarmists the result of this 80 ppm increase—presumably another 0.6°C of warming—would be taking us that much closer to disaster.

CO_2 warming falls exponentially

Or maybe not. If the CO_2-saturation effect is correct (meaning that the laws of physics haven't changed to accommodate climate alarmism), an additional 80 ppm of CO_2 would produce not 0.6°C of additional warming but 0.3°C. Another addition of 80 ppm would produce 0.15°C of additional warming. And so on.

Table 6.1: The carbon-dioxide saturation effect

Year	Added CO2 in ppm	Total CO2 in air 1900=290	°C temp. increase/80 ppm	Cumulative increase °C
1900-2000	80	370	0.6000	0.6000
2000-2100	160	450	0.3000	0.9000
2100-2200	240	530	0.1500	1.0500
2200-2300	320	610	0.0750	1.1250
2300-2400	400	690	0.0375	1.1625
2400-2500	480	770	0.0188	1.1813
2500-2600	560	850	0.0094	1.1906
2600-2700	640	930	0.0047	1.1953
2700-2800	720	1010	0.0023	1.1977
2800-2900	800	1090	0.0012	1.1988
2900-3000	880	1170	0.0006	1.1994

Table 6.1 works this out as an Excel spreadsheet up to the year 3000. (If you have Excel skills, you can easily calculate the progression for yourself—this book prides itself on its interactive, do-it-yourself approach!)

By the year 2600, with CO2 levels at a whopping ("catastrophic"!) 850 ppm, the cumulative yearly increase in CO2-induced warming shown in the fourth column is a tiny fraction above zero and the overall warming increase for our planet is 1.2°C, at least as far as CO2 is concerned. That's it. 1.2°C. Surely we can handle 1.2°C (or thereabouts) of warming!

Figure 6.1 shows what the carbon-saturation effect looks like as a graph (using a slightly different exponential figure)[4]:

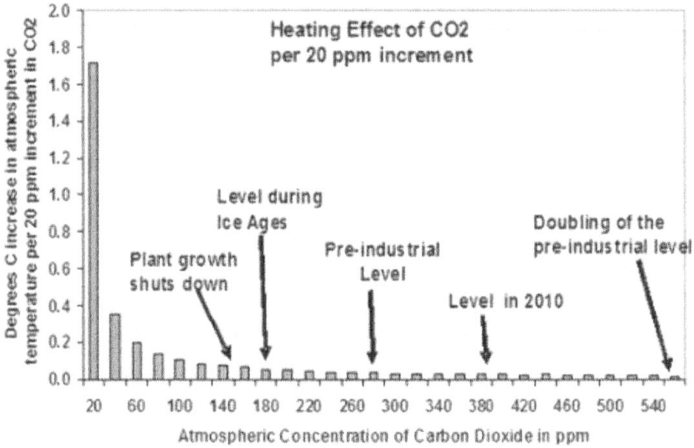

Figure 6.1: Carbon dioxide saturation effect. The temperatures shown by the bars need to be added to get the total temperature increase at any given CO2 level. Source: David Archibald

As you can see, by the label "Level in 2010" most of the warming effect of CO2 *has already occurred.* It's worth noting that plant growth shuts down at 150 ppm and we reached a low of 180 ppm during the most recent ice age. We were within 30 ppm of a dead planet—*that* is worth getting alarmed about!

To be clear: the carbon-saturation effect notwithstanding, the global temperature may well go above the magic number of 1.5°C from pre-industrial times (meaning before 1850).

But if that happens, it's because other warming factors than CO2 are involved, such as the Tonga volcanic eruption of water vapor, or the Sun, or the warming and cooling cycles (oscillations) of the ocean currents, or the effect of clouds, or El Niño cycles, or any of a hundred other factors that affect weather and climate *as well as,* and usually much more powerfully than, CO2 levels.

It's likely we will add 80 ppm or more of CO2 in the 21st century, Net Zero by 2050 or not. *But how much CO2 we put out really doesn't matter at this stage* because, at 425 ppm, the CO2 band in the atmosphere is almost certainly already "saturated," or very close.

This means any additional warming from additional CO2 will be minimal, despite what we're told by alarmist climate scientists. This is why, as we saw in Chapter 4, warming in the late 20th and 21st centuries has stalled between El Niños.

Real World physics versus Alarmist Wonderland Physics

It's worth noting that the CO2 band-saturation effect is in *complete opposition* to the theory that CO2 is the ultimate "control knob" of climate (lots more CO2 equals lots more warming without limit). Either one theory is right or the other, but they can't both be right.

Our best bet is to go with actual Real World Physics, rather than Alarmist Wonderland Physics (for a discussion on *why* Real World Physics and Wonderland Alarmist Physics are different, see Chapter 9 on "post-normal" science). To make carbon-dioxide increases "scary," alarmists have to resort to bringing in "positive feedbacks" and "tipping points" that may or may not exist (and where are the "negative feedbacks" in alarmist theory? Best not to ask....).

Meanwhile, back in Real World Physics, the warming effect of increased carbon dioxide at this point is minimal. If alarmists want to scare people about "global warming" they should find another bogeyman than carbon dioxide.

Now, I know all this sounds unbelievable, especially considering the hullaballoo from the IPCC and the alarmist climate-science establishment that we need to *drastically* cut CO2 emissions (and, effectively, wreck Western-style civilization) to keep the temperature increase at or below 1.5°C (a magical figure more-or-less picked out of a hat).

Fortunately, you don't have to believe me on this; the carbon-dioxide saturation effect is well-known to science and mentioned in many reputable sources, including physics textbooks and even some climate textbooks. For example, geologist John Bluemle writes:

> Atmospheric physics is quite clear that increasing CO2 concentrations increases temperature. The best way to demonstrate this is to model the temperature of the atmosphere with all the CO2 removed. It is very cold. Then, increase CO2 by small increments and plot the graph of temperature increase. *It is very rapid initially and then flattens out. Doubling CO2 from today's concentration, holding all other parameters constant, has a "negligible" effect.*[5] [emphasis added]

Climate scientist William F. Ruddiman is an alarmist when it comes to the pernicious warming effects of carbon dioxide.[6] However, even Ruddiman accepts the idea of CO2 band saturation, although he allows for more warming from added CO2 than my Excel table, writing:

> Earth's temperature reacts strongly to small changes in CO_2 values at the lower end of the range (less than 200 ppm), *but changes much less at the high end of the range* (greater than 800 ppm)."[7] [emphasis added]

At a CO2 level of 1,000 ppm, Ruddiman estimates, the Earth's average temperature would top out at 20°C, about 5°C higher than the 2022 average of 15°C,[8] and enough, he says, to melt the Arctic ice cap.[9]

Or not, because Ruddiman's "saturation level" for CO2 is higher than many other scientists'. For example, a recent paper suggests that the carbon band is saturated at 300 ppm, with a doubling producing warming of only 0.5°C. The author writes:

> Based on new radiative-transfer numerical evaluations, we reconsider an argument ... that saturation of the absorption of infrared radiation by carbon dioxide in the atmosphere sets in as soon as the relative concentration of carbon dioxide exceeds a lower limit of approximately 300 ppm. ... We find an equilibrium climate sensitivity (temperature increase due to doubling of atmospheric *CO_2* concentration) of about 0.50°*C*.[10]

Translation: At 300 ppm a doubling of CO_2 to 600 ppm will produce only 0.5°C of warming, more or less what is shown in Table 6.1. This means that CO_2 is not, and cannot be, the "control knob" of climate.

Whatever the saturation threshold (300 ppm sounds about right but it could be 200 ppm, too), if carbon-dioxide saturation is a real thing—based on actual, like, you know, Real World Physics—we don't have to worry about *uncontrolled* Thermageddon turning us all into crispy critters, and eco-alarmists should find another reason to terrify us into meek submission.

IPCC doesn't mention carbon band saturation

This failure to support climate alarmist theory is perhaps why *none of the IPCC's six reports directly mention the carbon-dioxide saturation effect.* CO_2 band saturation kind of spoils the alarmist party, so the IPCC's strategy may be: let's not mention CO_2 band saturation and hope no one notices.

Or perhaps the saturation effect is built into the alarmist computer models, but with added "positive feedbacks" and "tipping points" to make the temperature increase alarming again.

But in this case, we have to wonder why the resulting "adjusted" computer algorithms continue to make temperature so "sensitive" to CO_2 emissions that *the vast majority of the climate models continue to overpredict warming*. Even when climate modelers reduce the "sensitivity" a bit, they still end up with model temperatures higher than the observed, "hindcasted" temperatures.

These model failures suggest, to the non-alarmist anyway, that maybe the climate *was* "sensitive" to CO_2 at one point, but this high "sensitivity" has been left behind. The flat-lined temperatures of recent decades (except during El Niño events), shown in Chapter 4, argue strongly that the warming effect of added CO_2 is pretty much played out.

Here's another wrinkle: Alarmists would have us drastically reduce CO_2 levels to keep "global warming" in check. So consider Table 6.1 again and imagine carbon-band saturation going *in reverse* as we try to reduce carbon emissions to reach Net Zero (assuming that carbon dioxide is currently closely correlated to temperature, which it isn't, but assuming).

As Table 6.1 shows, an 80 ppm reduction in CO_2 from, say, 880 ppm to 800 ppm will reduce warming by a tiny fraction of a degree Celsius. The next 80-ppm reduction will reduce warming by a slightly larger but still tiny amount. The CO_2 reductions would have to be huge, say, down to 250 ppm, to have any marked effect on temperature.

At that point it's possible that Western industrial/technological civilization as we know it would be in ruins from lack of fossil fuels and also possible that a new glacial cycle would be on its way. A new glaciation really would bring down civilization and be far, far worse than 1.5°C or a bit more of warming.

El Niños, not CO_2, raise global temperature

To review: As noted in Chapter 4, CO_2 levels are rising steadily but the temperature is not. From 1990 to 2023, the factor bringing up overall average global temperature appears to be the El Niño cycle (you'll find background on the cause of the El Niño cycle in **Sidebar 3** following Chapter 4). For the last 70 years, outside of the El Niño years the temperature is basically flat or only rising slightly.

Here's the key take-away from this chapter: A flat-lining of temperature is *what you would expect if the warming effect of carbon dioxide was rapidly diminishing as CO_2 levels rise* (i.e., the CO_2 band in the atmosphere is saturated, or very close). This flat-lining is the *opposite* of what alarmist climate scientists expected, based on the "control knob" theory; as we'll see in Chapter 7, climate alarmists expected steadily rising temperatures.

And this means that attempts to drastically reduce CO_2 levels, like Net Zero by 2050, will have almost no effect on warming, but will have a drastic effect on our fossil-fuel-based economy. We'll discuss the economic effects of climate alarmism in Chapter 13 but first, in Chapter 7, let's look at the alarmist response to this inconvenient refusal of the planet to warm.

Notes for Chapter 6

---

1 IPCC, AR3 (2001). See https://archive.ipcc.ch/ipccreports/tar/wg1/index.php?idp=5

2 See, for example, Charles W. Rice & Richard Nelson, "What is Carbon and the Carbon Cycle?", Table 1. Montana State University. Available online as a PDF

3 Zeke Hausfather, "Analysis: Why scientists think 100% of global warming is due to humans." *Carbon Brief*, Dec. 13, 2017.

4 David Archer, "The Logarithmic Effect of Carbon Dioxide." *Watts Up With That*, March 8, 2010. Available online. Graph based on calculations by Willis Eschenbach.

5 John Bluemle, "Some thoughts on climate change." *North Dakota State Geological Survey Newsletter,* Vol. 28 (2001), pp. 1-2. Quoted in S. Fred Singer and Dennis T. Avery, *Unstoppable Global Warming: Every 1,500 Years*. Lanham, MD: Rowman & Littlefield, 2007, p. 179.

6 Ruddiman even wrote a book, *Plows, Plagues and Petroleum* (D.H. Freeman, 2001), arguing that the Little Ice Age (1350-1850) was caused by the Black Death—so many people died that the drop in global carbon dioxide from human activities, such as breathing and farming (we hadn't invented industry yet), cooled the planet (p. 100). Similarly, he argues that prehistoric agriculture, and particularly rice cultivation in China, produced enough carbon dioxide and methane, and therefore warming, to delay the start of the next glaciation, which, if past interglacials averaging 15,000 years are any measure, should be well on its way by now (p. 171). He acknowledges that this hypothesis is "provocative and controversial," but his argument is interesting: if human carbon-producing activities kept the glaciers at bay, isn't that a Good Thing? Also, CO_2 levels were struggling upward from their ice-age levels of 200 ppm, when the warming effect of CO_2 increases was stronger than today's levels.

7 William F. Ruddiman, *Earth's Climate: Past and Future*. New York: D.H. Freeman, 2001, p., p. 134. This is one of my favorite climate textbooks.

8 NOAA estimates the average global temperature in 2022 was about 15°C. See https://www.climate.gov/news-features/understanding-climate/climate-change-global-temperature.

[9] Ruddiman, *Earth's Climate*, p. 135.

[10] Dieter Schildkneckt, "Saturation of the Infrared Absorption by Carbon Dioxide in the Atmosphere." *International Journal of Modern Physics B*, Aug. 5, 2020. Available online at Cornell University site https://arxiv.org/abs/2004.00708.

Chapter 7: Why isn't the world warming fast enough?

Now, here, you see, it takes all the running you can do to keep in the same place.

For climate alarmists, "The Science" is "settled," "certain," and beyond question. After all, "The Science" is based on a "consensus" of more than 2,000 scientists who contribute to the IPCC's reports. Us non-scientists—politicians, the media and the public—are supposed to accept "The Science" without question.

And indeed, climate scientists are highly resistant to publicly debating, or even defending, "The Science" in public with critics, even though debate and criticism are the life-blood of science, because agreeing to debate gives "credibility" to climate critics (the "denial industry").

Do I appear to be exaggerating? Alas, no. For example, climate modeler Andrew Weaver, one of the leading figures in climate alarmism in Canada, has written:

> There is *no ... debate* [over human carbon emissions as the primary cause of global warming] in the atmospheric or climate scientific community, and ... making the public believe that such a debate exists is precisely the goal of the denial industry.
>
> There is a clear, well-understood reason for global warming: It is largely because of carbon-dioxide emissions from the combustion of fossil fuels. ... *Scientific debate over global warming would therefore imply uncertainty. ... There is no such debate in the atmospheric and climate scientific community.*[1] [emphasis added]

Al Gore himself has written, in his usual convoluted syntax:

> If, *when the remaining unknowns about the environmental challenge enter the public debate*, they are presented as signs that the crisis may not be real after all, it undermines the effort to build a solid base of public support for the difficult actions we must soon take.[2] [emphasis added]

As we've seen, and will continue to see, there are still plenty of "remaining unknowns" in climate science, but for Gore and Weaver and many other believers in climate doom ahead, we should just ignore these pesky little unknowns.

Instead, we should start—now! immediately!—to spend trillions of dollars on "green," "sustainable" forms of energy (the staggering costs of Net Zero are discussed in Chapter 14).

And if these measures destroy our fossil-fuel-based industrial/technological civilization, a civilization that, for all its undoubted problems, is also the source of myriad benefits (including electricity, modern medicine, running water, painless dentistry, and so on)—well, that's a small price to pay for "saving the planet," whatever that means.

Climate models: much more warming than reality

Given all this certitude, you'd expect that the climate computer models, which aim to predict future warming based on increased CO_2 levels, would be remarkably accurate. That is, if alarmist theory is correct, you'd expect the planet to co-operate and warm the amount the models predict.

Alas for alarmist theory, this has not been the case. As we saw in Chapter 4, the planet has stubbornly refused to warm as much as the vast majority of the computer models predict. We know this by looking back and comparing the model predictions with what has *actually occurred* over the past few decades ("hindcasting").

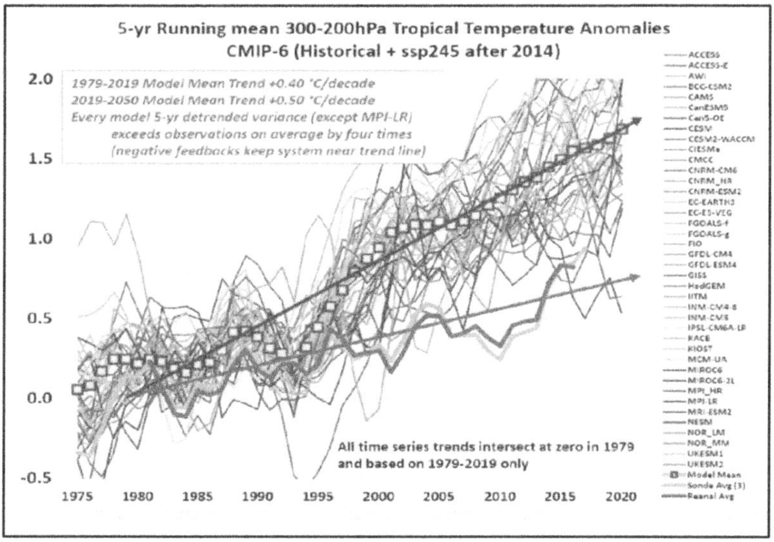

Figure 7.1: Computer models predict far more warming (top 'spaghetti' lines) than has actually occurred (lower lines). Source: John Christy.

Even worse, as we saw in Chapter 4, when "hindcasted" the climate model predictions are not just a bit off but wildly off by orders of magnitude (see Figure 7.1, repeated from Chapter 4[3]: the thick top line is the average of the computer predictions; the lower lines are actual measured temperatures plus the straight-line average). If climate was as "certain" and "settled" as alarmists would have the public believe, wouldn't the models be more accurate?

This large gap between theory and reality means that the relationship between carbon dioxide and temperature is, perhaps, not as completely understood as alarmist climate science would have us believe (there are, perhaps, many "remaining unknowns"!). Maybe we aren't doomed to die a fiery death after all!

As touched on in Chapter 4, the problem of not enough warming first became public in the mid-2000s: not only was the predicted scary warming not happening, but warming had *stalled*! OMG! The planet wasn't warming as much as the models said it should have! There was, as you can imagine, considerable alarm among alarmist climate scientists at this failure of Empirical Reality to match "The Science."

The Climategate emails

How do we know the scientists were seriously concerned? Because, in 2009, a whistleblower within Britain's Climatic Research Unit (CRU) released thousands of emails containing the uncensored thoughts of a cohort of leading alarmist climate scientists. (CRU is part of the British Meteorological Centre's Hadley Institute, which produces the HadCrut temperature data.)

This release of documents became known as Climategate, and Climategate painted a rather unflattering picture of climate scientists and of the state of alarmist climate science.[4]

Many of these emails dealt with two major alarmist concerns:

1. One concern was the fact that in the early 2000s the planet not only wasn't warming as it should, which was heresy enough, but that the planet might even be—brace yourself!—cooling!
2. A second concern was how to deal with those pesky climate skeptics, some of whom were fellow climate scientists on the non-alarmist track, who impertinently pointed out this failure to warm. It's one thing to dismiss non-scientist skeptics as "deniers" (even though many non-scientists are well-informed on climate issues). It's not so easy to dismiss critics who are climate scientists themselves, not to mention scientists in other disciplines who, presumably, might actually have some professional scientific expertise (and if you believe the hype about 97 per cent of climate scientists supporting apocalyptic human-caused global warming, I have a bridge to sell you[5]).

Let's start with the first topic: not enough warming. We'll cover dealing with the bothersome climate critics ("deniers") in Chapter 9, which details where climate science took a major step away from accepted science into a swamp called "post-normal" science.

So: why wasn't the planet warming as the models said it should?

For a start, and repeating some of the material from Chapter 4, Figure 7.2 shows the "warming" between 2000-2014 based on the land-sea readings of the Hadley Centre (HadCrut) and Goddard Institute for Space Studies (GISS), and satellite readings by the University of Alabama at Huntsville (UAH) and Remote Sensing Systems (RSS). What do we see in Figure 7.2? All four readings are pretty much *flat* (UAH is actually *decreasing*!).

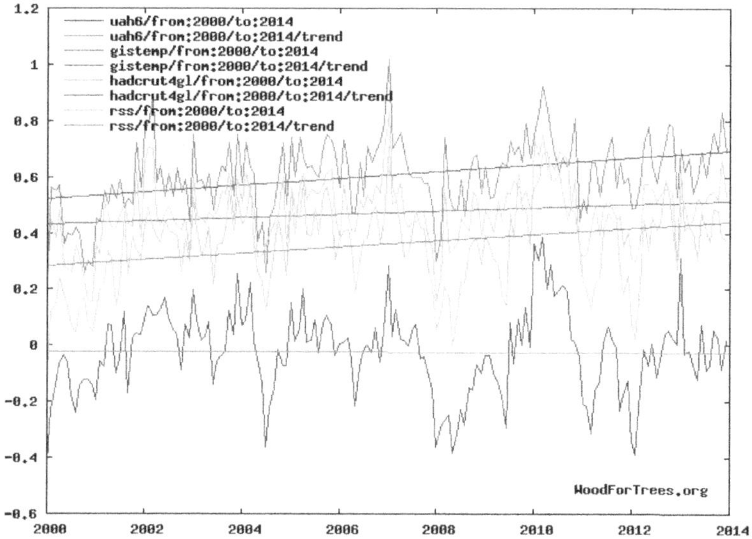

Figure 7.2: Temperature readings for 2000-2014 from the four major climate-monitoring agencies; they are pretty much flatline, so where's the "accelerated" warming? Source: Woodfortrees.org

In July 5, 2005, well into this non-warming trend, Phil Jones, head of the Climatic Research Unit, wrote the following in an email that was part of the Climategate trove:

> The scientific community *would come down on me in no uncertain terms if I said the world had cooled from 1998.* OK it has but it is only seven years of data and it isn't statistically significant.[6] [emphasis added]

But wasn't the world supposed to be *warming*? And, horrors, it wasn't, at least from 1998-2005.[7] Fortunately, Jones notes, this apparent non-warming wasn't "statistically significant," a convenient mathematical fig leaf for alarmists when Reality isn't behaving as it should.

Another interesting wrinkle in Jones's email: that the "scientific community" would "come down" on him "in no uncertain terms" if he publicly admitted the world had actually cooled a bit since 1998.

Aren't scientists supposed to report the facts without fear or favor (a.k.a., "objectivity")? Apparently not, at least when it comes to climate science; it's much more important to stick to the doomsday "message." We'll say more about the alarmists' need to maintain scientific "consensus," even at the cost of the truth, in Chapter 9.

As it turned out, to everyone's consternation, and as we saw in Chapter 4, the temperature stayed pretty much flat not just for 10 years but for *14 years*, from 2000 to 2014, until a very large El Niño came up in 2015-16 to (partially) save the day.[8] I say partially because after the El Niño the temperatures went flat again.

'Hiatus'? Or just a 'temporary slowdown'?

Alarmist climate scientists called this embarrassing 14 years the "pause" or "hiatus,"[9] with the clear implication to us ignorant non-scientific folks that this non-warming was only a *temporary* lull: again, nothing to see here folks, move along! They promised that the CO2 planet-burning would be back with a vengeance once the "hiatus" was over!

More recently, some alarmist climate scientists have simply declared that the "hiatus" never happened and that the appearance of flat-lined temperatures were caused by a teeny glitch in how the temperature data was handled; at worst, there might have been a "temporary slowdown."[10]

This process is commonly known, again, as torturing the data to fit the theory.

Unfortunately for alarmist climate science, this "temporary slowdown" also seems to have occurred from 2017 to 2023 (we saw this in Chapter 4 but it's worth repeating). Even NOAA, which showed more warming from 2000-2014 than the four Woodfortrees climate agencies, has admitted that global land and sea temperatures from 2015-2022 have been *flat* and even *falling* (see Figure 7.3).

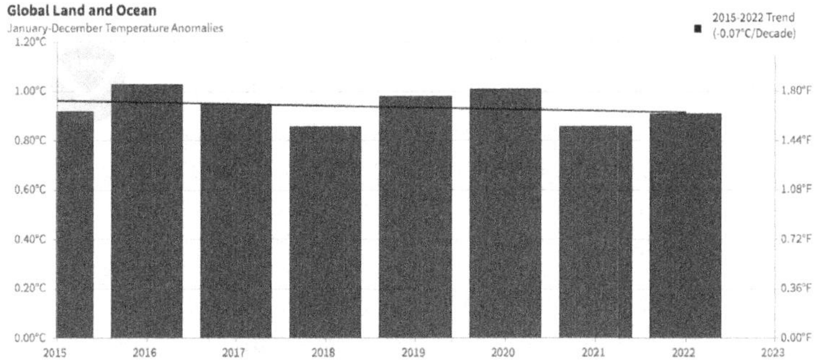

Figure 7.3: Global land and sea winter temperatures from 2015-2022. Temperature falling! Source: NOAA, "Global Time Series" temperature

Far from warming, the NOAA graph shows that from 2015 to 2022, in the January-February months, the planet *cooled* at a rate of minus-0.07°C per decade in the winter (January-December timeline). The summer timeline (August-July) showed zero warming. Flat. Nada. Zip. Note that these aren't actual temperatures but anomalies, the deviation from temperatures between 1901-2000. Still, not warming compared to the 20th century!

This is an odd turn of events if, as the IPCC and NOAA insist, increased CO2 levels should be causing considerable warming. It's almost as if warming had, like, you know, slowed down or even stopped....

But nothing daunted, NOAA doggedly continues to fill its website with items aimed at creating climate alarm ("Hottest June on record!", etc.) while other alarmists desperately try to conjure away this latest six- or seven-year "temporary slowdown."

Alarmists fight claims of 'cooling' from 2015-2023

For example, Reuters' "Fact Check" webpage laments that climate deniers have deviously leapt onto the NOAA 2015-2023 data to claim that the globe is "cooling." Reuters notes, helpfully:

> Data from short periods of time showing cooling temperatures cannot be extrapolated to longer-term trends and do not prove that long-term global warming is a hoax. [**Author's note**: just for the record, extrapolation of short-term trends to longer trends is a scientific sin that climate alarmists have committed quite frequently, so this an odd defence.] Moreover, ... the warming effect of carbon emissions is cumulative and would not be expected to track with yearly fluctuations.
>
> When better corrections for various sources of bias are applied to the data ... the so-called global warming hiatus vanishes—and in fact ... *global warming may have sped up*. That won't startle some scientists, who say the "hiatus" was always a misnomer. 'There is no hiatus or pause," says climate scientist Michael Mann of Pennsylvania State University, who prefers the term 'temporary slowdown'."[11] [emphasis added]

For alarmists, then, "global warming" is "speeding up" when the temperature—which is surely the most basic indicator of global warming or cooling—shows no warming whatsoever. And isn't a "temporary slowdown" the same thing as an "hiatus" or "pause"? You can't make this stuff up.

Some of the scientists interviewed by Reuters in the same article suggest that this recent multi-year temperature flat-line is due to a larger than usual La Niña, the cooler side of the warm 2015-16 El Niño cycle. Reuters writes:

> A large El Niño event followed by a La Niña can "lead to a temporary 'pause' in global temperatures over timescales of a decade or so," climate scientist Zeke Hausfather is quoted as saying. This is "what we are now seeing after the 2015/2016 super El Niño event."[12]

Or maybe—and this interpretation fits the climate facts better than the idea of a "temporary slowdown" to explain the 2000-2014 and 2015-2022 lack of warming—the *El Niño cycle* is acting more like the "control knob" of climate at the moment than CO2, given that the CO2 atmospheric band is almost certainly "saturated," as noted in Chapter 6.

The El Niño "control knob" theory seems far more "harmonious" with the actual non-warming data than the alarmist theory, at least for those not blinded by alarmist ideology.

Just to repeat: In the time periods 2000-2014 and 2015-2023, the four climate-monitoring agencies show little or no warming. So that's *20 years of no or little warming within the past 23 years*—another fact that doesn't "harmonize" with the alarmist claim that the planet is warming at a steady (and even "speeded up") pace.

Nor, as we saw in Chapter 4, was there much warming from 1979-1994. Nor from 1947-1979. It was only the massive 1998 El Niño that brought the temperature average up for the last half of the 20th century and the 2015-16 El Niño that brought up 21st century temperatures.

Canadian cities: No warming for more than a century

Another indicator of global warming, or lack of it, is the temperatures of Canada's cities for the past century.

Here, once again in do-it-yourself mode, there's an online gadget that can give you this information. It's called Weatherstats.ca and shows the temperatures over time for many Canadian cities and towns, drawing on data from Environment and Climate Change Canada.[13]

The details on how to use the gadget are on the Weatherstats website, or you can find detailed instructions on the Climate Realists of British Columbia website (climaterealists.ca/cities)[14], but what you'll get initially is a temperature chart over 25 years for the city you chose that looks like Figure 7.4, the data for my home town, Victoria, from 1996-2021.

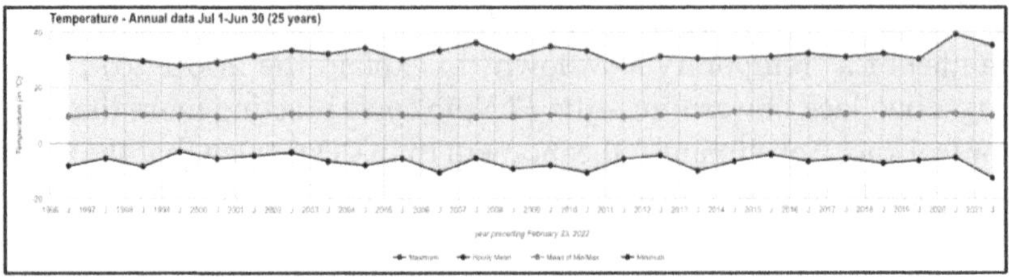

Figure 7.4: Average temperature in Victoria, B.C., from 1996-2021. Source: Weatherstats.ca

The top line shows average high temperatures; the bottom line shows average low temperatures; and the middle line is the average of the high and low temperatures.

As you can see, the average line is *dead flat*, during a 25-year time span between 1996-2021 when, if we believe alarmists, the temperature should have been rising alarmingly like the blade of a "hockey stick." Astonishingly, you can go back in 25-year increments as far back as 1913 for Victoria and the average line continues to be flat.

Victoria has a rather mild climate all year round, so I tried Weatherstats.ca to see if Regina, Sask., with its hot summers and very cold winters, showed any scary warming over more than a century. The result: flat temperatures all the way back to 1913 (Figure 7.5 shows temperatures from 1994-2019 but the line is Prairie-flat back to 1913).

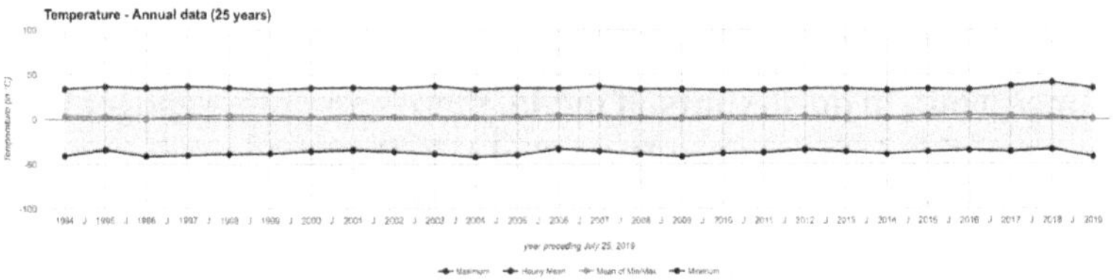

Figure 7.5: Regina temperatures from 1994-2019. Source: Weatherstats.ca

You might wonder how this can be, given what we've been told by the alarmists. The answer is that most of the alarmist climate graphs, with the scary ups and downs of temperature (the temperature anomalies), represent only a small range of temperature, usually 1°C or less.

This means a small amount of warming looks like a Himalayan series of peaks and valleys, as you can see in Figure 7.6, which is the temperature rise and fall (mostly fall) from 2002-2014 using two satellite monitoring agencies, RSS and UAH.

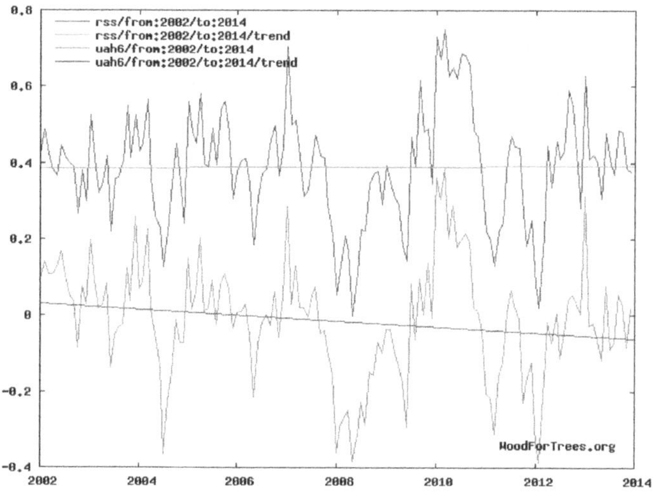

Figure 7.6: Average temperatures from 2002-2014 for RSS and UAH climate-monitoring agencies. Source: Woodfortrees.org

The jagged temperature line looks like a roller coaster because the temperature range (left axis) is only 1.2°C—less "climate change" than you'd experience in a single day.

Pull back to the much wider temperature highs and lows of the Weatherstats.ca graphs—Regina's chart has a range of 100°C, for example—and the temperature line *flattens out*. From this zoomed-out perspective there's no jutting "hockey stick" blade, no "accelerated warming," in any Canadian city you'd care to try. Which I hope you will—it's an eye-opener!

'We cannot account for the lack of warming'

But getting back to the non-warming in the early part of the 21st century: Kevin Trenberth, a lead IPCC author, lamented in one Climategate email: "The fact is that *we cannot account for the lack of warming at the moment* and it's a travesty that we can't."[15] [emphasis added] Does this sound like The Science is "settled"?

Similarly, environmental writer Mark Lynas, author of the hyper-alarmist book *Six Degrees: Our Future on a Hotter Planet*, wrote in 2008:

> Although CO2 levels in the atmosphere are increasing each year, *no one ever argued that temperatures would do likewise*. Why? Because the planet's atmosphere is a chaotic system, which expresses a great deal of inter-annual variability *due to the interplay of many complex and interconnected variables*. Some years are warmer and cooler than others.[16] [emphasis added]

No one ever argued that just because CO2 levels were up, temperatures would do likewise? Is he kidding? Isn't that temperature tracks CO2 levels *exactly* what climate alarmism argues, time after time? And isn't that what the computer models predict? And doesn't this "great deal of inter-annual variability *due to the interplay of many complex and interconnected variables*" rather argue against the theory that CO2, on its little lonesome, is the climate "control knob"?

To answer the rhetorical question: That CO2 and temperature go up (or down) pretty much in lock-step, with CO2 in the lead, is *exactly* the argument that Al Gore makes (implicitly) with his huge 650,000-year wall chart. This is *exactly* the argument that the IPCC computer modelers make (and are then wildly off when their models are "hindcasted" against Reality).

Also, "some years are warmer or cooler than others" (a rather fatuous remark, don't you think?). Cooler for *14 years*? While CO2 steadily rises? Doesn't Lynas's explanation seem like something of a last-ditch attempt to put a finger in the leaky dike of alarmist ideology?

In full damage control, Tim Flannery, an Australian arch-alarmist, observed in 2009:

> In the last few years, where there hasn't been a continuation of that warming trend, *we don't understand all of the factors that create Earth's climate, so there are some things we don't understand,* that's what the [Climategate] scientists were emailing about. ... These people work with models, computer modeling. *When the computer modeling and the real world data disagree you have a problem.*[17] [emphasis added]

Yes, you do have a problem, including what might happen if the public, media and public twigged to the fact that maybe "we don't understand all of the factors that create Earth's climate."

If this is so then, hmm, maybe The Science isn't as "settled" and "certain" as climate alarmists (like Lynas and Flannery, among many others) would have us all believe as they lead us down the (very expensive) garden path of "green" energy to prevent a supposed climate apocalypse that, so far, refuses to make an appearance except in the climate models.

Horrors! Maybe CO2 *doesn't* control the climate! Maybe cutting CO2 emissions, at huge cost to our way of life (as we'll see in Chapter 14), won't make any difference to the global temperature! We can't let the public believe that!

Ignore the data, go with the models

But rather than seriously acknowledging the problem with the 2000-2014 "hiatus"/"temporary slowdown" (like, maybe our theory is wrong), the alarmist community circled the wagons, choosing to stick with the theory and ignore the reality.

One climatologist, Chris Folland of the Hadley Climate Centre, stated the stick-to-the-story, eyes-wide-shut strategy with refreshing honesty: "The data don't matter.... *We're not basing our recommendations on the data.* We're basing them on the climate models."[18] [emphasis added]

Problem solved! Just go with the models, even if the temperature data disagrees, while covering up the non-warming (and even a bit of cooling) as best you can (just a "temporary slowdown," folks! Nothing to see here!) and pray the globe heats up again (which, thankfully for the alarmist theory, it did in 2023, but not, sadly, due to CO2, as we saw in Chapter 5).

This doubling down by prominent climate scientists and their representatives was pretty much standard procedure during the non-warming of the early 2000s. For example, Rajendra Pachauri, the then-head of the IPCC (although himself an engineer not a climate scientist), visited Australia in 2008 and told an audience there: "We're at a stage where warming is taking place *at a much faster rate*."[19] [emphasis added]

Really? When the temperature evidence so clearly showed, even in 2008, that warming had stalled (as Figure 7.2 reveals)? You can't make this stuff up. And if the *head* of the IPCC can't get his facts straight, either through ignorance or as a deliberate attempt to deceive the public, why would we trust the rest of his organization?

Similarly, Vicky Pope, a spokesperson for the Hadley Meteorological Centre, declared in February 2010 (take note of the month): "If anything, the world is warming *even more quickly* than we had thought."[20] Eventually Reality caught up with Pope and, in November, *a mere nine months later*, she told reporters almost the *complete opposite*: "There's a very clear warming trend but it's *not as rapid as it was before*."[21] [emphasis added] Presumably, then, the planet was *not* warming "even more quickly" just months earlier.

It's curious, isn't it, that someone who works for a meteorological agency wasn't aware of this less-clear, "less-rapid" warming trend (actually a *non-warming* trend, but why quibble over terminology?) a mere nine months earlier in February, when (for Pope) the world was warming "even more quickly." Once again, as so often, you can't make this stuff up.

Climate modeler Andrew Weaver simply ignored the temperature data when he told a Victoria, B.C., magazine interviewer in 2011: "The allegation that annual global mean temperatures stopped increasing during the past decade *has no basis in reality.*"[22] [emphasis added]

Unpacking this a bit: Climate alarmists point to *rising* (even better, "record") temperatures as a sign of uncontrolled global warming not just in the future but *now*, in the present. But they *ignore* or *discount* temperatures if they are *flat or falling* because the temperature data isn't giving the right "message."

Can you have 'global warming' if planet isn't warming?

Sure, there are other measures of warming (or cooling), like sea-level rise, etc. And yet, isn't temperature *the most basic indicator of "global warming,"* if serious global warming is, in fact, occurring? I mean, can you have "global warming" if the planet isn't actually, like, *warming*? On the Alarmist Wonderland side of Alice's mirror, yes; in the Real World side, no.

If you find this alarmist version of reality bizarre, it could be because you are a bit of a scientific traditionalist who believes, as I do, that scientists are supposed to base their theories on *actual empirical data*, like, say, temperature readings, not computer models.

This empirical approach is how science is practiced in most other disciplines, to great success. It's what I've tried to do in this book. You don't have to *believe* me—I've tried to offer not just actual facts, but show you where you can find these temperature facts for yourself (e.g., Woodfortrees.org, Weatherstats.ca, even NOAA).

As a champion of traditional science, physicist Richard Feynman wrote: "If it [the theory] disagrees with experiment it is wrong. In that simple statement is the key to science."[23] Except in Climate Science Wonderland—for example, compare Feynman's approach with that of Chris Folland, noted above, that "We're not basing our recommendations on the data. We're basing them on the climate models."

Feynman's view is supported by Thomas Huxley's comment: "The great tragedy of science—the slaying of a beautiful hypothesis by an ugly fact." It is what Aristotle meant when he suggested that strong theories were "harmonious." Strong theories don't have "ugly facts," or not many. But alarmist climate science has scores of these ugly facts! It not only lacks Aristotelian "harmony," it is positively incoherent at times.

Taking the argument a bit further: which approach strikes you as having more scientific integrity? Acknowledging to other scientists and the public that an "ugly fact" or two has popped up, thereby putting your theory at risk? Or do you carry on, regardless of "ugly facts," with your head in the sand?

The honest approach is what philosopher of science Karl Popper advocated as the basis of good science when he wrote:

> If you are interested in the problem which I tried to solve by my tentative assertion, you may help me by *criticizing it as severely as you can*; and if you can design some experimental test which you think might refute my assertion, *I shall gladly, and to the best of my powers, help you to refute it.*[24] [emphasis added]

As we've seen in the Climategate emails, alarmist climate scientists *actively resist* this openness to criticism and refutation.

Policy based on dicey computer models

And yet alarmist climate scientists are continuing to make anti-fossil-fuel recommendations to politicians, based on dicey computer models that in "hindcast" have not reflected the climate reality.

And, alarmingly, "progressive" politicians like Canadian Prime Minister Justin Trudeau and U.S. President Joe Biden and many European leaders are actually *acting upon* these recommendations, in the process doing serious damage to their economies and to the workers and families who depend on those economies.

This doesn't mean the temperature won't suddenly decide to zoom upward at some point, as it has in 2023. It just means the planet hasn't warmed very much for the last 30 years, and really since at least 1950, *except when an El Niño spike intervenes*. Shouldn't we be worried by alarmist climate scientists' disconnect from reality?

The many "ugly facts" in alarmist climate science raise the question: are we dealing with real, empirically based *science* here? Or some sort of religious faith, in this case, faith in computer models rather than a religious book? Climate alarmism has too many parallels with end-of-the-world religious fanaticism to be a coincidence.

So, to repeat the question: is climate science an *actual science*? We'll examine that question in Chapter 9. But first, let's look at the kind of horrible world we could be facing if CO2 rises much further and the global temperature gets any hotter.

Notes for Chapter 7

1 Andrew Weaver, *Keeping Our Cool: Canada in a Warming World*. Toronto: Viking Canada, 2008, pp. 22-23.

2 Al Gore. *Earth in the Balance: Ecology and the Human Spirit*. Toronto: Penguin Books, 1993 (1992), p. 39.

3 We saw this figure in Chapter 4 as well. It is available at Clintel. https://clintel.org/new-presentation-by-john-christy-models-for-ar6-still-fail-to-reproduce-trends-in-tropical-troposphere.

4 For a selection of these emails, see "Climate cuttings 33" at the Bishop Hill blog site. The URL is http://bishophill.squarespace.com/blog/2009/11/20 /climate-cuttings-33.html. For a more complete discussion of the Climategate scandal, as well as the text of most of the emails, search for John Costella, "The Climategate Emails."

5 See, for example, Lawrence Solomon, "97% cooked stats." *Financial Post*, Jan. 3, 2011. Available online.

6 Again, for details, see the Costella reference.

7 You may be wondering why I didn't start the graph in 1998. This is to avoid charges of "cherry picking" the data. 1998 was an El Niño year, so temperatures were very high. If 1998 is the starting year for the graph, then naturally the trend will be down (i.e., "cooling"). Starting after the El

Niño year gives a clearer picture of just how little warming there actually was. I do this as well in a graph of the non-warming between 2017-2023.

8 Peter Ferara, "The period of no global warming will soon be longer than the period of actual global warming." *Forbes*, Feb. 24, 2014.

9 For example, see "Hiatus confirmed as scientists finally admit climate model fail." *CFACT*, Feb. 26, 2016. Available online. Also, Robert K. Kaufmann, et al., "Reconciling anthropogenic climate change with observed temperature 1998–2008," *PNAS* (Proceedings of the National Association of Scientists), July 19, 2011. The paper notes: ""Data for global surface temperature indicate little warming between 1998 and 2008. Furthermore, global surface temperature declines 0.2 °C between 2005 and 2008."

10 Carolyn Gramling, "Much-touted global warming pause never happened," *Science*, June 4, 2015.

11 Reuters Fact Check, "Fact Check—Eight cooler years cannot be extrapolated to draw conclusions on long-term global warming." *Reuters*, Feb. 3, 2023. Available online.

12 Reuters, Ibid.

13 Ross McKitrick, "Reality check—there is no 'climate emergency' in Vancouver." *Vancouver Sun,* July 23, 2019. See also https://climaterealists.ca/cities for a complete discussion of how to Weatherstats.ca.

14 Paul MacRae, "Temperatures in Canadian cities: Where's the scary warming?" *Climate Realists of British Columbia*, May 9, 2022. URL: climaterealists.ca/cities.

15 Kevin Trenberth to Michael Mann, Oct. 12, 2009. See note 3, above, Costella, "Climategate Emails."

16 Mark Lynas, "Has global warming really stopped?" *New Statesman*, Jan. 14, 2008.

17 "Flannery defends scientists in leaked emails row." *ABC News online*, Nov. 24, 2009.

18 Quoted in Patrick J. Michaels, *Sound and Fury: The Science and Politics of Global Warming*. Washington: Cato Institute, 1992, p. 83.

19 Michael Duffy, "Truly inconvenient truths about climate change being ignored." *Sydney Morning Herald,* Nov. 8, 2008. For an in-depth (and disturbing) analysis of Pachauri's deliberate misrepresentation of the climate facts, see Christopher Monckton's "Open Letter to Chairman Pachauri," Science and Public Policy, Dec. 16, 2009.

20 Jonathan Leake, "World may not be warming, say scientists." Times Online, Feb. 14, 2010. Of course, there are other measures of warming than temperature, as Kevin Trenberth, a lead IPCC author, notes in the same Leake article. "It's not just temperature rises that tell us the world is warming," he said. "We also have physical changes like the fact that sea levels have risen around five inches since 1972, the Arctic icecap has declined by 40% and snow cover in the northern hemisphere has declined." But surely these changes are dependent on temperature.

21 Alex Morales, "World May Post Warmest Year as U.K. Met Office Adjusts Past Decade of Data." *Bloomberg*, Nov. 26, 2010.

22 Amy Reiswig, "Tunnelling through the Wall of Hate." *Focus*, July/August 2011, pp. 34-35.

23 Richard Feynman, *The Character of Natural Law.* Cambridge, MA: MIT Press, 1965, p. 156.

24 Karl Popper, *Conjectures and Refutations: The Growth of Scientific Knowledge.* New York: Routledge & Kegan Paul, 2007 (1963), p. 35.

Sidebar 4: Canadian city 'warming' using weatherstats.ca

Weatherstats.ca is a website that offers data on weather conditions in Canadian cities now and over time since 1913. The site draws its data from Environment and Climate Change Canada and, assuming Environment Canada isn't tampering with the data to show more warming, should be fairly reliable.

Astonishingly, as economist Ross McKitrick pointed out in a 2019 *Vancouver Sun* article,[1] the site shows that for more than a century the average temperatures for most Canadian cities have been pretty much *flat*. Yes, *flat*.

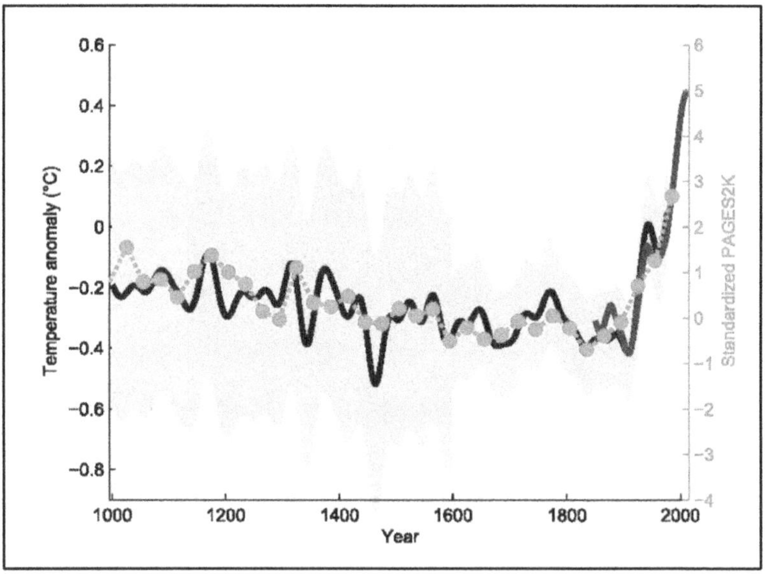

Figure SB4.1: The IPCC's "hockey stick" graph, which appears to show massive "warming" in the 20th century

If you are looking in the city temperature data for the IPCC's iconic "hockey stick," which shows temperatures flat or falling since 1000 AD, conveniently erasing the Medieval Warm Period, ending with massive temperature zooms in recent times (Figure SB4.1), you won't find it. And if *Canadian* cities aren't warming, it's likely that cities in other parts of the world aren't, either, or not very much.

Fortunately, as with Woodfortrees (**Sidebar 2**), you can check the data for yourself on the weatherstats.ca website. Here's how.

Figure SB4.2: Cities of Canada

Go to weatherstats.ca. You'll be presented with a map of Canada (Figure SB4.2); the circled numbers represent Canadian cities and towns. Click on the circle with the city you want (zooming in will reveal more cities in any given area).

Let's select Victoria, B.C., as an example since this city is a hotbed of climate-alarmist activism (and because I live here). You will get the page shown in Figure SB4.3.

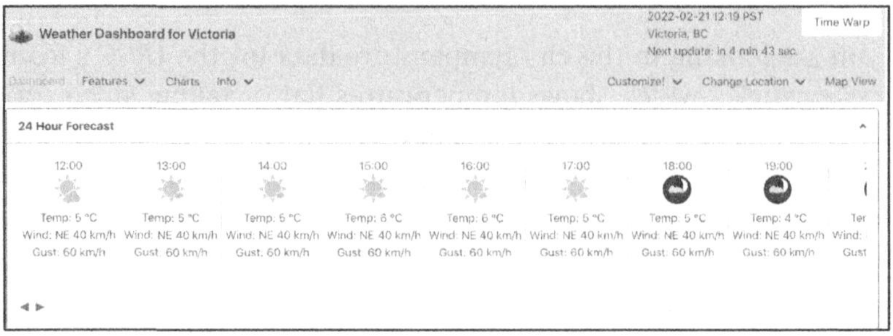

Figure SB4.3: Dashboard for Victoria

We want the historical weather data for Victoria, so click on "Charts" on the top menu of the webpage. You'll get the page shown in Figure SB4.4, below.

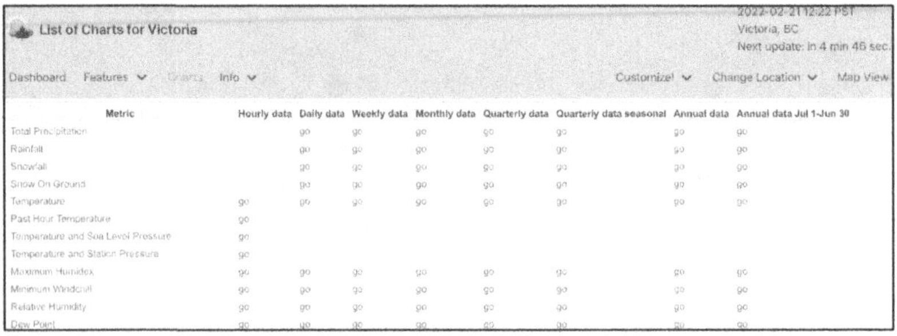

Figure SB4.4: View of "Charts" with city weather data

One of the "go" options is "Temperature" (fifth choice from the top in this example); this should tell us how drastically Victoria has heated up over the past few decades.

So click on "go" under "Annual data Jul1-June30" in the "Temperature" line, which starts and ends the data series in mid-summer. (If you click on "Annual data" the 25-year series begins and ends in mid-winter, which makes the green line dip down a bit at the end.) You'll get Figure SB4.5.

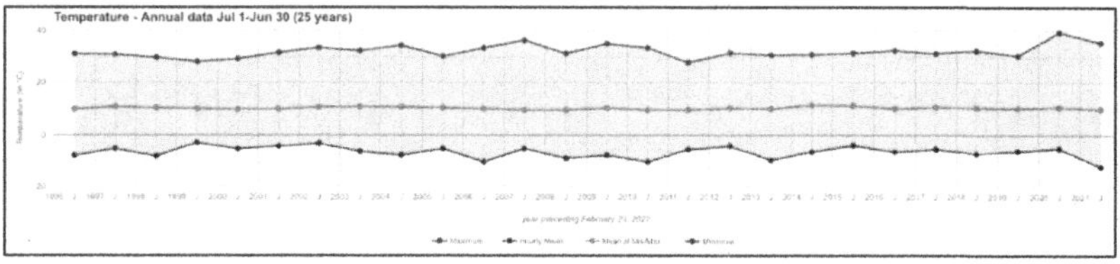

Figure SB4.5: Temperature changes for Victoria from 1996 to 2021

The **top line** shows the average high temperatures for this 25-year-period; the **bottom line** is the average low; the **middle line** is the average of the high and low temperatures. As you can see, for Victoria the average line is pretty much dead flat *for 25 years*, from 1996 to 2021. This flat temperature isn't what we'd been led to expect by the climate alarmists!

How far back in time does Victoria's flat-lined temperature go? The website has data back to summer 1913. You can view dates earlier than 1996 in 25-year blocks by clicking on the "Time Warp" button at the top right of the heading (see Figure SB4.3), then changing the number in the "Years:" box (see Figure SB4.6).

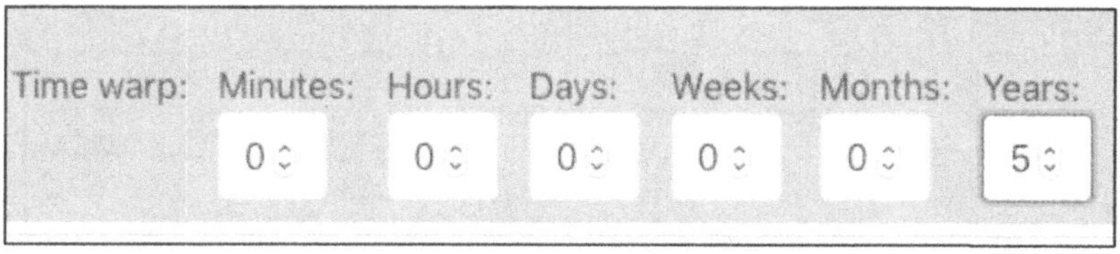

Figure SB4.6: Choose an earlier time period

The box has an up-down arrow to take you back one year at a time, but you can also input any number to go back that number of years (I put in "25" years but only the "5" shows for some reason).

Figure SB4.7 shows the average Victoria temperatures from 1971 to 1996. Once again, we have dead flat.

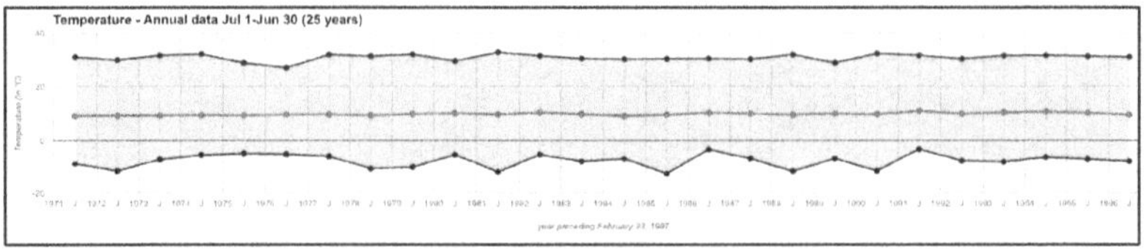

Figure SB4.7: Average Victoria temperatures from 1971-1996

If we go back "50" years, to 1946-1971, we get Figure SB4.8. Maybe the warming started then. But nope; still flat.

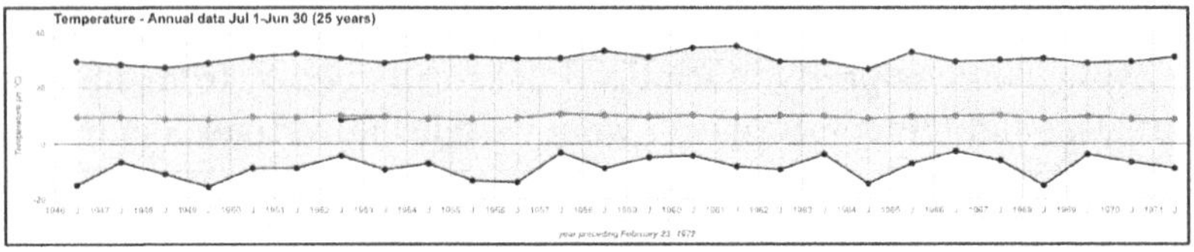

Figure 2SB4.8: Average Victoria temperatures from 1946-1972

Let's type in "75" to go back another 25 years (1921-1946), as shown in Figure SB4.9. Flat.

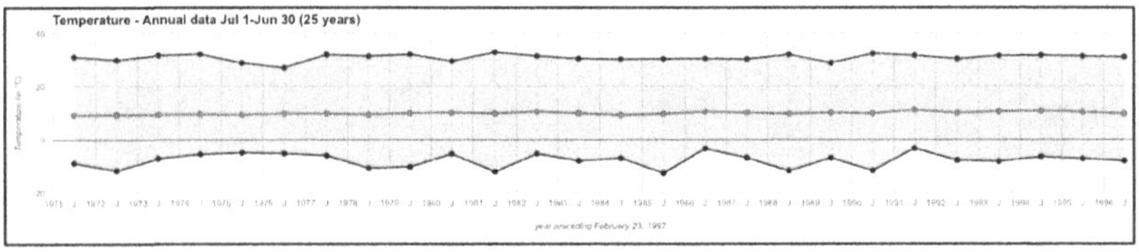

Figure SB4.9: Average temperatures 1921-1946

And Figure SB4.10 shows the profile from 1913, the farthest back the site goes, to 1938 (I used the "down" arrow under "Year" to go back from "1921" to "1913"). Once again, virtually dead flat.

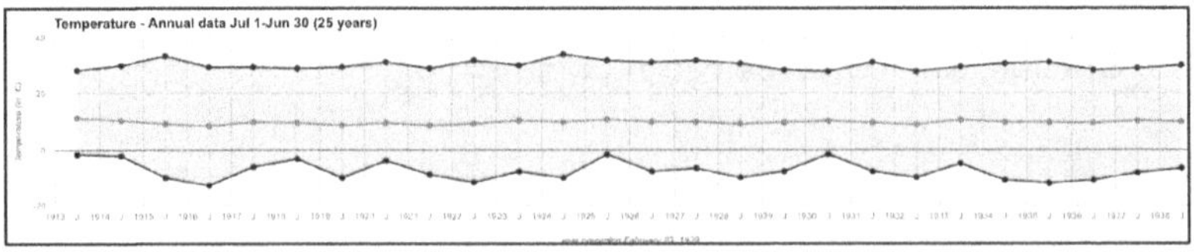

Figure SB4.10: Victoria temperatures from 1913-1938

Give it try for any Canadian city on the map. You'll find the same thing: there is no massive warming, as we've been told is occurring. There's no warming at all!

How can this be? Again, it's because the "hockey stick" type of graph (SB4.1) uses a 1°-2°C temperature range, so small changes look big. The city data for Victoria, for example, uses plus-40°C to minus-20°C, a range of 60°C. On this scale, warming is virtually non-existent.

For more details on how to use weatherstats.ca, see the Climate Realists of British Columbia website at climaterealists.ca/cities.

Chapter 8: Carbon dioxide—friend or foe?

It sounded an excellent plan, no doubt, and very neatly and simply arranged; the only difficulty was, that she had not the smallest idea how to set about it.

If you go to the alarmist *Carbon Brief* website and search for "Boris," you'll find an article entitled "Revealed: The 11 slides that finally convinced Boris Johnson about global warming," dated Feb. 1, 2022. Johnson, then Britain's prime minister, had been skeptical about climate-change alarmism, but these 11 slides reportedly changed his mind.[1]

One slide in particular probably caught Johnson's attention (see Figure 8.1). The map shows much of the northern hemisphere as *red-hot* (in black and white, the darker grey shades are the flaming red bits in the color version of the slide). Sizzling! Scary!

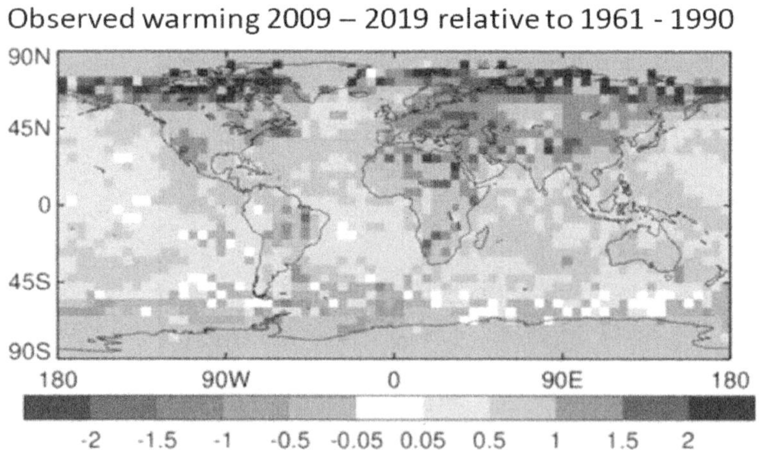

Figure 8.1: British Meteorological Office map showing scary warming

It's easy to imagine that these are the *real* temperatures when they're not; they're departures (anomalies) from the 1961-1990 average, thirty years when the planet was a bit cooler than today.

The Arctic and northern regions in general look like they're on fire when, in fact, yes, the Arctic regions are warming more quickly than the temperate and tropical regions, but the North is still plenty cold in winter.

Also, if you look at the Arctic on an actual globe, you'll realize this region is relatively small compared to the rest of the planet—just under three per cent of the world's total surface area.[2]

In short, the graph uses scary red colors and a Mercator map format that expands the size of the Arctic to terrify the unwary, and this image (and several others) apparently helped persuade Boris Johnson to board the alarmist climate train.

A greener planet

Figure 8.2 shows another way of looking at our warming world that doesn't show degrees of red but degrees of *green* (grey here).

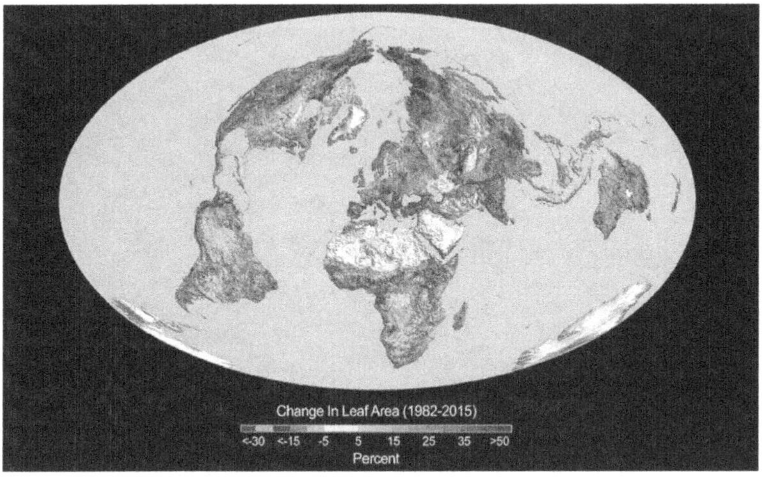

Figure 8.2: Change in leaf area from 1982-2015. Source: NASA

This image also has a graduated colored scale across the bottom but, in this case, the "extreme" right end shows increased *greenery*. Curiously, the image comes from NASA, no friend of climate realism. The NASA web posting writes:

> From a quarter to half of Earth's vegetated lands have shown significant greening over the last 35 years *largely due to rising levels of atmospheric carbon dioxide*, according to a new study published in the journal *Nature Climate Change* on April 25 [2016].[3] [emphasis added]

What? Increased carbon dioxide *not* destroying the Earth but actually making it greener! And how much greener? NASA says *5 per cent greener* in the last 20 years, the equivalent of the Amazon rainforest![4] Who could have predicted this greening of the world in the midst of a global climate catastrophe?

Well, among others, greenhouse growers could have predicted this. They know that carbon dioxide is a fertilizer for plants—indeed, CO_2 is essential for plant life. At a CO_2 level of 150 ppm or lower, plants can no longer survive,[5] which means most forms of life, including us, could not survive.

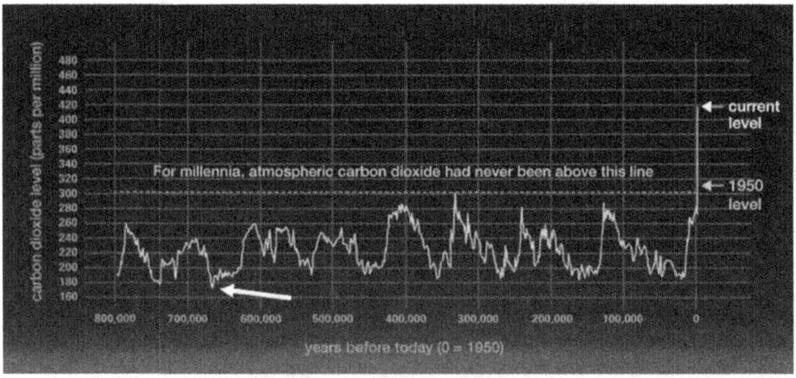

Figure 8.3: CO_2 levels over the last 800,000 years. CO_2 levels came close to the doomsday 150 ppm about 650,000 years ago (white arrow). Source: NASA

As you can see from Figure 8.3 (previous page), also from NASA, our planet almost reached this do-or-die CO2 point about 650,000 years ago, when we were at a close-to-rock-bottom 180 ppm (white arrow on graph).[6]

And, as NASA notes, for much of the last 800,000 years, CO2 levels averaged a low of 200 ppm, which is a bit too close to 150 ppm for comfort, don't you think?

Curiously, NASA offers this graph as proof that we are experiencing *too much* CO2 these days. Puzzling. You'd think we'd want to get as far above the 150 ppm mark as possible, that current CO2 levels aren't *too high* but *too low*! And you'd think Greens would happy about CO2 plant-enrichment since it produces more of their favorite color, but they're not. There's just no pleasing alarmists.

How CO_2 enhances plant growth

Anyway, to repeat, one of the side effects of, for NASA, *too much* atmospheric carbon dioxide (weird) is plant growth that is not only enhanced but more drought-resistant. Here's how it works.

Plants have little openings on the bottom of their leaves called stomata. These stomata open up to trap carbon molecules from the atmosphere, which plants need to grow their stems and leaves, etc. At the same time, whenever the stomata open, the plant loses moisture, which can be a problem in drier regions. So it's a balancing act for the plant: CO2 or moisture?

When CO2 levels are low and carbon molecules are not plentiful in the air, plants need to leave the stomata open longer to get the carbon they need, losing more moisture. But as CO2 levels increase (take a bow, humanity!) there are more carbon molecules flying around and it's easier for plant stomata to scoop them up. That means better plant growth.

Because the stomata don't have to stay open as long to get the carbon the plant needs, there is also less water loss, which means more drought-resistant plants.[7] That's why the Earth is greening, especially in the semi-arid regions.

Again, based on this greening you'd think more CO_2 in the atmosphere would be a Good Thing, even for alarmists, but apparently not.

Even the IPCC agrees that more CO_2 equals more greening. In its sixth report (2021) it writes: "Increasing atmospheric CO_2 concentration enhances leaf photosynthesis and drives a partial closure of leaf stomata, leading to higher water-use efficiency at the leaf canopy and ecosystem scales."[8]

Here's a fact that may shock you (and should certainly shock climate alarmists): as noted above, greenhouse growers actually *add* CO_2 to their greenhouses to reach levels of 1,000 ppm or more. That's more than twice the current level and well into doomsday territory for climate alarmists. Who knew?

Here's what an Ontario government website for greenhouse farmers has to say about enhanced (but, for alarmists, "catastrophic") CO_2 levels:

> *CO_2 increases productivity through improved plant growth and vigor.* Some ways in which productivity is increased by CO_2 include earlier flowering, higher fruit yields, reduced bud abortion in roses, improved stem strength and flower size. *Growers should regard CO_2 as a nutrient.*
>
> For the majority of greenhouse crops, net photosynthesis increases as CO_2 levels increase from 340 to 1,000 ppm (parts per million). Most crops show that for any given level of photosynthetically active radiation (PAR), *increasing the CO_2 level to 1,000 ppm will increase the photosynthesis by about 50% over ambient CO_2 levels.*[9] [emphasis added]

Pretty good for a trace gas that is, for Australian ultra-alarmist Tim Flannery, "choking the atmosphere," that is "global warming pollution" for Al Gore, and just plain "pollutant" for the U.S. Environmental Protection Agency (EPA).[10]

On the negative side, alarmists point out that these enhanced superplants might not be quite as nutritious (have slightly less protein and nitrogen) as their less-developed cousins.[11] More food but it might be slightly less nutritious. For climate alarmists, there's always a worm in the apple....

'Oblivion'? Or 'paradise lost'?

As I was doing research for my first book, *False Alarm*, I was extensively immersed in (and astonished by) the contradictory, incoherent, non-"harmonious" nature of climate alarmism.

On the one hand, the alarmist climate scientists whose books and articles I read were "alarmed" at the rising CO2 level. At the same time, they often described earlier geological times, *when temperatures were several degrees higher and the CO2 level was at least two or three times higher than today's*, as "tropical paradises."

For example, Tim Flannery is the author of the best-selling ultra-alarmist *The Weather Makers,* which predicts global catastrophe from warming. However, in his very readable 2002 book *The Eternal Frontier: An Ecological History of North America and its Peoples,* Flannery had this to say about the Eocene in North America, 50 million years ago, when CO2 levels were as high as 3,000 ppm or as low as 1,000 ppm (depending on which sources you consult):

> When Earth is warm (in greenhouse mode)—as it was around 50 million years ago—North America is a *verdant and productive land.* Almost all of its 24 million square kilometers, from Ellesmere Island in the north to Panama in the south, *is covered in luxuriant vegetation.*[12] [emphasis added]

The subheading for the section of *The Eternal Frontier* quoted above? "In which America becomes a tropical paradise." CO2 levels were at least 1,000 ppm (and probably higher) and yet North America was "verdant and productive." This coming from a climate ultra-alarmist. As so often with alarmist climate science, you can't make this stuff up!

Similarly, paleo-climatologist Donald R. Prothero has written a very readable book about the last 50 million years, including the Eocene. He calls the Eocene a "lush, tropical world" despite CO2 levels well above 1,600 ppm and temperatures several degrees (Prothero suggests 10°C) warmer than today. The title of the book? *Eocene-Oligocene Transition: Paradise Lost.*[13]

If a world up to 10°C warmer than today, with CO_2 levels at least three times higher, could be "paradise lost," does it make any sense at all to believe, given today's much lower temperatures and CO_2 levels, not to mention being in the midst of an Ice Age, that we are heading inevitably toward some sort of *global-warming* catastrophe?

And yet this catastrophic future is what alarmists believe in, or at least what they would like the public to believe in. For example, during the 2007 climate conference in Bali, then-UN secretary-general Ban Ki-Moon said in a speech: "Today, we are at a crossroads, one path leading towards a comprehensive new climate agreement, and the other towards *oblivion*. The choice is clear."[14] [emphasis added]

"Oblivion." If the temperature went up 2°C? See "tropical paradise," above.

More recently, in March 2023, the IPCC released its Sixth Assessment Synthesis Report (part of AR6). In announcing the report, UN Secretary-General Antonio Guterres didn't quite say "oblivion," but he came close:

> Humanity is on thin ice — and that ice is melting fast. As today's report of the Intergovernmental Panel on Climate Change (IPCC) details, *humans are responsible for virtually all global heating over the last 200 years*. The rate of temperature rise in the last half century *is the highest in 2,000 years*. Concentrations of carbon dioxide are at their highest in at least 2 million years. *The climate time-bomb is ticking*.[15] [emphasis added]

And when the "climate time-bomb" goes off, presumably we'll experience "oblivion." "Tropical paradise" isn't, apparently, on the menu. Nor on the menu, for Guterres, is looking at the *actual temperature data*, which does *not* show a terrifying increase in the rate of temperature rise (see Vicky Pope, above).

Guterres' speech contains an alarmist grocery list of economic measures developed nations need to take if we're going to save the planet, including no new coal mines and phasing out of all coal use by 2040, a halt to oil and gas expansion and shutdown of oil and gas by 2050, and so on.

I'll discuss these Net Zero by 2050 economic measures in Chapter 14, but just as an aside: While the developed nations of North America and Europe are cutting their own energy throats, China is currently building two coal-fired power plants *a week*, while flooding the Western market with solar panels, as it seeks energy security (and world hegemony) through fossil fuels, not wind and solar.[16]

"Sustainable energy" is for the suckers in the West. You can't make this stuff up.

Media should fact-check doomsday claims

Of course, the mass media dutifully reported Guterres' speech with the usual doom-laden headlines, as if his ideas made economic and environmental sense, without doing any actual, you know, *investigation* into whether this claim of a ticking "climate time bomb" (or, for that matter, "oblivion") is in any way factual or even possible.

Why should we believe the fear-mongering of Ban Ki-Moon or Antonio Guterres or any of the climate alarmists when we know, *as a fact*, that in the geological past the Earth's temperatures were *much warmer* and CO2 levels *much higher* than today, and that Earth was—even according to alarmist writers like Flannery and Prothero—a "tropical paradise"?

The idea that humanity faces an *existential* challenge—a threat to our very existence—from "climate change" is absurd and not supported by any empirical evidence whatsoever; however, there's lots of empirical evidence (again, see "tropical paradise," above) that we *aren't* heading to climate disaster.

Surely even alarmist climate scientists must be aware of the two UN leaders' disconnect with reality. But if so, why didn't these scientists try to correct Ban Ki-Moon's delusory 2007 comment? Why don't they point out the wild exaggeration in Guterres' 2023 remarks? This silence speaks volumes about the integrity of alarmist climate scientists.

Even the ultra-alarmist climate scientist Stephen Schneider has admitted: "In the sense that it [climate change] will be an urgent catastrophe which will eliminate all life, that kind of statement is absolutely unwarranted."[17]

Is there, then, any reason to believe that "oblivion" awaits humanity if we choose to much more *slowly and gradually* adapt our economic and social systems to whatever "global warming" occurs, using the many technological tools at our disposal? Instead, we've accepted the UN's crash anti-fossil-fuel program, which really could mean "oblivion" for millions, perhaps billions, of people as their nations' economies shrink and, in some cases, collapse.

Wouldn't it make more sense to turn a deaf ear to the alarmists and enjoy the extra greening that more carbon dioxide is bringing to our planet—to the point of becoming again, perhaps, a "tropical paradise"?

Notes for Chapter 8

---

1 You'll find the article on the *Carbon Brief* website using search terms "Johnson change climate policy slides."

2 The area of the Arctic is 5.5 million square miles; the area of the entire Earth is 197 million square miles. Doing the math, the Arctic represents 2.7 per cent of the Earth's surface, but looks much larger in a Mercator map projections. Sources: For area of Arctic, NOAA, "Frequently asked questions about the Arctic." For area of the Earth, One World Nations Online, "The Earth." Both are online.

3 NASA, "Carbon dioxide fertilization greening Earth, study finds." April 26, 2016.

4 Magnus Aschan, NASA, "The Earth is greener now than it was 20 years ago." March 5, 2021. The subheading reads: "The Earth has become five percent greener in 20 years. In total, the increase in leaf area over the past two decades corresponds to an area as large as the Amazon rainforests."

5 Ministry of Agriculture, Government of Manitoba, "Greenhouse CO2 supplement." The site offers advice to greenhouse gardeners and notes: "At 100 PPM of CO_2 the rate of photosynthesis would be stopped completely. At 150 PPM the plants begin to respire, and photosynthesis is stopped. At this low level the plant will no longer be able to obtain CO_2 from the atmosphere and photosynthesis is restricted. The plant will eventually use all of the CO_2 present, photosynthesis will stop and the plant will die."

6 NASA, "Graphic: The relentless rise of CO2." No date.

[7] Physicist Freeman Dyson describes the process in his essay "The Greenhouse Effect: An Alternative View," in his book *From Eros to Gaia*, London, Penguin Books, 1993. It is online as a chapter of the anthology *Eyewitness to Science*. The URL is https://archive.org/stream/B-001-024-525-ALL/B-001-024-525_djvu.txt.

[8] IPCC, Climate Change 2021: The Physical Science Basis. Chapter 5, Cross-Chapter Box 5.1, p. 697. Available online. Earlier reports have noted that additional CO2 can increase plant growth by up to 33 per cent. See IPCC 2001, Chapter 3, Section 3.2.2.4, p. 195.

[9] "Supplemental carbon dioxide in greenhouses." Ontario Government. Dec., 2002.

[10] For details, see Wikipedia, "Regulation of greenhouse gases under the Clean Air Act." Basically, perhaps because it hasn't taken complete leave of reason, the EPA doesn't call carbon dioxide a "pollutant" directly; instead, it got the courts to agree that CO2 fell under its Clean Air Act, which is aimed at "pollutants."

[11] Cell Press, "Increasing levels of CO2 results in less nutritious crops." *Science Daily*, March 27, 2023.

[12] Tim Flannery, *The Eternal Frontier: An Ecological History of North America and its Peoples*. London: Vintage, 2002, p. 84.

[13] Donald R. Prothero, *Eocene-Oligocene Transition: Paradise Lost*. New York: Columbia Univ. Press, 1994, pp. 22, 238, 35.

[14] United Nations News, "UN Climate Change Conference represents crossroads, Secretary-General says." Dec. 7, 2007.

[15] United Nations, "Secretary-General Calls on States to Tackle Climate Change 'Time Bomb' through New Solidarity Pact, Acceleration Agenda, at Launch of Intergovernmental Panel Report, March 20, 2023.

[16] Robert Lyman, "China is winning the climate policy game." *National Post*, March 28, 2023.

[17] Stephen Schneider, *Science as a Contact Sport*. Washington, DC, National Geographic, 2009, p. 94.

Chapter 9: Real Science vs. 'Post-Normal' Science

"If there's no meaning in it," said the King, "that saves a world of trouble, you know, as we needn't try to find any."

Reading between the lines, dear reader, you may have guessed that I have little respect for alarmist climate scientists and their claims of catastrophe ahead based on a "Science" that is supposedly "settled" and "certain" and beyond debate. I want to assure you that this scorn for alarmist climate science does not, emphatically not, mean that I do not respect science in general. I absolutely do.

I believe science is humanity's all-time most important invention. That's why I hate, with a white-hot passion, to see science and just basic truth misused and even degraded, as it is so regularly by alarmist climate scientists and environmentalists.

When I began serious research into the apocalyptic claims of alarmist climate science for my first book, *False Alarm,* back in 2007, I was continually astonished at the weakness of the evidence for these claims.

Worse, I was appalled at how often alarmist climate scientists failed to follow even the most basic standards of good science, including objectivity, reliance on *empirical* evidence (not computer models), scrupulous regard for the truth, rigorous testing, replication of results by other researchers, and openness to criticism and debate, among others.

But then, of course, if you are saving the planet you shouldn't have to follow the rules of normal science. It's okay to bend the rules, even if you are a "scientist," because the end justifies the means.

In Chapter 7 I mentioned that this alarmist lack of openness to criticism appeared often as a second major theme in the 2009 Climategate emails.

These emails included extensive discussions on how to suppress those annoying skeptics who just don't get it (perhaps because the skeptics prefer to follow the empirical data rather than the computer models).

In the Climategate emails are many examples of alarmist climate scientists violating the basic scientific norm of openness. For example:

- Phil Jones, head of the Climatic Research Unit, writing of two papers by skeptics that actually got published in a peer-reviewed journal: "I can't see either of these papers being in the next IPCC report. Kevin [Trenberth] and I will keep them out somehow—*even if we have to redefine what the peer-review literature is.*"[1] [emphasis added]
- Another email has climatologist Tom Wigley discussing proposals to get the editorial board of the peer-reviewed scientific journal *Climate Research* to resign because the journal had been publishing articles by skeptics.[2]

In a similar vein, although not part of Climategate, Richard Somerville of the Scripps Institute of Oceanography has written:

> People who are not experts, who are not trained and experienced in this field [climatology], who do not do research and publish it following standard scientific practice, are not doing science.[3]

In other words, for alarmists, we, the public, have *no right* to comment on climate matters, even if the anti-fossil-fuel, "green" policies that Somerville's "experts" propose to save the world will severely damage or even eliminate our livelihoods (as we'll see in frightening detail in Chapters 14 and 15).

For example, Alberta oil-fields workers might have a stake in the dramatic downsizing of their industry that the alarmist climate scientists are encouraging. So might drivers forced to buy expensive electric cars, and farmers being told they can't use nitrogen fertilizers and forced to reduce their cattle herds, as is happening right now in some countries.

Should non-scientists have *no* role in assessing a science if that science threatens their livelihoods in the name of some greater good that exists only in computer models?

Philosopher Karl Popper would not have agreed with that position. While lay people should respect the views of experts, he noted that "a non-scientist might criticize science for failing to abide by its own publicly avowed standards."[4]

Alarmist climate science violates scientific norms

I have argued so far in this book, and will further in the next chapter in more detail, that alarmist climate scientists have strayed far, far away from science's "publicly avowed standards." Instead, they are leading society into an economic and social catastrophe, while claiming, on very slight evidence, to be saving us from a climate catastrophe.

In fact, I believe this is the first time in history that *scientists* have proposed policies that will make most of us worse off, rather than better off. Worse, these policies are based on an apocalyptic CO2 "control knob" theory that, as the flat-lined 2000-2014 and 2015-2022 temperature data shows, doesn't have a lot of actual real-world empirical support.

(As an aside and just to repeat: Temperatures have risen in 2023, largely due to the massive undersea eruption in the Pacific in 2022, which added tons of water vapor to the atmosphere—a 10-15 per cent increase, as we saw in Chapter 5. Water vapor holds heat; more water vapor holds more heat.

(When we look at the actual temperature data, we see that the "record" July 3, 2023, temperature of 17.01°C, for example, was only 0.09°C warmer than the previous "record," in August 2016, of 16.92°C.[5] Are we supposed to cower in terror because the global temperature has gone up a whopping 0.09°C, one-tenth of a degree Celsius, an increase so small that the human body can't even detect it?

(Nor do "record temperatures" in 2023, or any other year, prove that reducing carbon emissions will somehow bring these temperatures down.

(Also, these "record" temperatures are based on instrument readings that began in the mid-1800s—so the "record" is only for the past 170 years or so. Could the planet have been warmer before that? Answer: Yes, repeatedly, e.g., the Medieval Warm Period. So are these "record" temperatures something to worry unduly about? Answer: No.)

Alarmist 'science' is 'post-normal' science

Overall, then, I was baffled as to how these violations of the usual scientific norms could have occurred in alarmist climate science.

I hadn't realized at the time, until I was informed by a colleague,[6] that alarmist climate scientists were, consciously and deliberately, *not* following the "publicly avowed standards" of science. Instead they were following a *different* type of science than traditional science, with *different* standards of honesty, integrity and evidence. They were following "post-normal" science.

Wikipedia describes "post-normal" science as follows:

> Post-normal science (PNS) ... is a problem-solving strategy appropriate when "facts [are] uncertain, values in dispute, stakes high and decisions urgent," conditions often present in policy-relevant research. In those situations, *PNS recommends suspending temporarily the traditional scientific ideal of truth, concentrating on quality as assessed by internal and extended peer communities.*[7] [emphasis added]

Although I hadn't been aware that "post-normal science" was A Thing, I'd actually been faced with it time and time again over the years as I researched alarmist writings about climate; I just didn't have the right label.

For example, in its 2001 report, the IPCC is quite explicit about its reliance on the post-normal approach, without using the actual term:

> We fully recognize that many of the evaluation statements we make contain *a degree of subjective scientific perception and may contain much "community" or "personal"*

> *knowledge*. For example, the very choice of model variables and model processes that are investigated are often based upon *the subjective judgment and experience of the modeling community*.[8] [emphasis added]

IPCC contributor Mike Hulme describes the resort to "post-normal" science this way:

> The problem is that in areas of science which are seeking to understand the behavior of large complex systems which can't be replicated in the lab, *it is very hard if not impossible to apply the scientific litmus test of falsification through experimentation*. And climate change is one such area of science.
>
> We have scientific theory, we have empirical observations. What we haven't got are lots of different Earths that can be experimented on in controlled conditions. *Virtual climates created inside computer models are the best we've got*.[9] [emphasis added]

He goes on:

> Climate scientists frequently have to reach their conclusions on the basis of the *partial, and sometimes poorly tested, evidence and models available to them*. And when their paymasters—elected (or non-elected) politicians—ask them for advice, as in the case of the IPCC, *opinion and belief become essential for interpreting facts and evidence*.
>
> Or rather, incomplete evidence and models have to be worked on using opinions and beliefs to reach considered judgments about what may be true.[10] [emphasis added]

Stephen Schneider put post-normal science this way: "We can't put the Earth in a test tube," which means "computer modeling is our *only available tool* to perform what-if experiments such as the human impact on the future."[11] [emphasis added]

Which means relying on "opinions and beliefs" rather than empirical facts to advise alarmist climate science's government "paymasters." And, as we all know, the one who pays the piper calls the tune....

'Post-normal' science relies on 'consensus'

To summarize, "post-normal" science operates on *subjective* judgments reached by a "consensus" of "experts" ("quality as assessed by internal and extended peer communities"), at the beck and call of "elected (or non-elected) politicians" (i.e., *politicized* science, see Chapter 11 for more details on the strong political bias of alarmist climate science).

It is also reliant on computer modelling rather than the traditional scientific norms of empirical evidence and rigorous testing of hypothesis. The comment by Chris Folland, noted in Chapter 7, sums up "post-normal" science quite well: "The data don't matter.... *We're not basing our recommendations on the data.* We're basing them on the climate models."[12] [emphasis added]

This is why the IPCC reports rely on a sliding scale of certitude with terms like "unlikely," "likely," "very likely," or "low confidence," "medium confidence," "high confidence," and so on, as in:

> Additional warming is projected to further amplify permafrost thawing and loss of seasonal snow cover, of land ice and of Arctic sea ice (*high confidence*). The Arctic is likely to be practically sea ice-free in September at least once before 2050 .... There is *low confidence* in the projected decrease of Antarctic sea ice.[13] [emphasis is the IPCC's]

We're not relying on empirical facts in "post-normal" alarmist climate science; we're primarily relying with computer-aided guesses about the future. And on the basis of these computer-aided guesses, politicians are expected to radically retool Western society (as we'll see in Chapters 15 and 16).

Testing is the gold standard of normal science

"Post-normal" science is, then, not the science of Einstein and Feynman, which is based largely on Karl Popper's "falsifiability" principle, the gold standard of modern science that is accepted by almost all scientific disciplines (at least in theory).

For Popper, science—meaning valid science—works as follows: scientists study phenomena, create an hypothesis that they think explains how the phenomena works, and then design *rigorous tests* of the hypothesis, usually by making predictions. If the theory is viable, the predictions will be accurate. If not viable, the predictions will fail.

The importance of testing as part of "normal" science can't be emphasized enough. For example, a first-year, university-level textbook on science notes: "The requirement that we be willing *to subject our explanations to experimental tests is the distinguishing feature of science.*"[14] [emphasis added]

Similarly, a book of readings in the philosophy of science states that one key criterion of "science" is testability: "If the propositions which express [scientific] claims *are not capable of tests, we cannot call these propositions true or false or even how to go about establishing their truth or falsity.*"[15] [emphasis added]

Note that, as Hulme and Schneider (among others) admit, alarmist climate science is "not capable of tests." *This suggests we should be highly skeptical in evaluating the reliability of the apocalyptic predictions of "post-normal" alarmist climate science.*

But also note: for Popper, scientific tests aren't designed to "prove" an hypothesis; they are designed to "disprove" or "falsify" it, in the same way that automobile makers give their cars "torture tests" on a race track. As Popper put it, the key question for a scientist wasn't "Am, I right?", but "Under what conditions would I admit that my theory is untenable?"[16]

(Just for fun, try asking a climate alarmist this question: "What evidence could I give you that would persuade you that your theory is wrong?" They won't answer, or will evade the question, because they consider their theory beyond question—indeed, they believe the theory *shouldn't* be questioned![17] An odd sort of science, this "post-normal" science, and more reminiscent of religious dogma than open intellectual enquiry.)

The "falsification" criterion means that a theory can never be finally, incontrovertibly, *proved* to be "true" ("settled"/"certain") for all time. There is always a chance that new, completely unanticipated data could disprove (falsify) the theory. Therefore, scientists should keep open minds. Or, as Nobel-prize-winning physicist Richard Feynman put it, "a scientist is never certain."[18]

Also, for Popper, a scientist tests a theory by making predictions, and the riskier the better. For example, in the climate context, that it will snow next winter is not a risky prediction; that it will *not* snow next winter, a prediction that climate alarmists have made (almost always wrongly) from time to time,[19] *is* risky and therefore more worthwhile, scientifically.

If risky predictions based on the theory are borne out again and again, then the theory likely has merit (although it may still be falsified in the future, so still not "certain"). If predictions consistently fail (as they have so often in alarmist climate models), then the theory is clearly shaky.

The key point here is that, for Popper and many traditional scientists, scientific tests may *refute* a theory, but they can never establish it as "certain" because, Popper argues, the whole function of the "true scientific attitude" is to *undermine* ideas that science is "settled" and "certain" and beyond question.

This attitude, for Popper, "is utterly different from the dogmatic attitude which constantly claimed to find 'verification' for its favorite theories."[20] As, for example, alarmist climate science does. Also, Popper notes, "A theory that is not refutable by any conceivable event is not scientific,"[21] which pretty much sums up alarmist climate science.

Science is open to criticism

In his most famous book, *The Open Society and Its Enemies*, Popper supports the idea of an "open" society, whose members "treat openness to criticism and change as a personal ethic and a civic duty."[22] [23]

Based on what we've seen so far, and especially in the Climategate emails, it's clear that *leading* alarmist climate scientists do not share this openness to criticism. Nor are they willing to offer any "conceivable event" that would falsify their theory.

Nor do they share Einstein's view, which he applied to his own theory of relativity:

> The chief attraction of the theory [of relativity] lies in its logical completeness. *If a single one of the conclusions drawn from it proves wrong, it must be given up*; to modify it without destroying the whole structure seems to be impossible.[24] [emphasis added]

Or Charles Darwin's attitude, in *Origin of Species,* toward his theory of evolution:

> I am well aware that scarcely a single point is discussed in this volume on which facts cannot be adduced, often apparently leading to conclusions directly opposite to those at which I have arrived. *A fair result can only be obtained by fully stating and balancing the facts and arguments on both sides of each question.*[25] [emphasis added]

This open-minded attitude couldn't be further from the closed-mind, "settled" and "certain" stance of alarmist climate science.

How different traditional science is, then, from the "post-normal" attitude of alarmist climate science, which seeks to shut down criticism, claims certainty, and is unable to perform falsifiable tests!

And based on this "post-normal," subjective, non-falsifiable approach to "science," we—the public, media and politicians—are being asked, to turn our lives upside down to fight a problem (catastrophic human-caused "global warming") that may exist only in computer models and alarmist imaginations.

That said, understanding that there is a "post-normal science" perspective based on "consensus" and intolerant of criticism cleared up much of the puzzlement I had about the failure of climate scientists to follow the traditional scientific method and norms.

It wasn't that they were blatantly or ignorantly flouting the norms of traditional science, as I believed (although they are flouting the traditional norms). They are just operating from the "post-normal" scientific paradigm rather than traditional science.

Climate-science groupthink

Not surprisingly, the "post-normal" emphasis on "consensus" makes alarmist climate scientists much more susceptible than most scientists to groupthink.

Groupthink, according to Yale psychologist Irving L. Janis and author of the 1982 book *Groupthink*, occurs when "subtle constraints ... *prevent a [group] member from fully exercising his critical powers* and *from openly expressing doubts* when most others in the group *appear to have reached a consensus*."[26] [emphasis added].

In groupthink, even if the group leader expresses an openness to new ideas, group members *value consensus more than critical thinking*; groups are thus led astray by excessive "concurrence-seeking behavior."[27] Groupthink is "consensus" on steroids

This definition of groupthink fits alarmist climate science to a "T". It's why climatologist Keith Briffa could write to fellow scientist Michael E. Mann (the inventor of the IPCC "hockey stick," which shows modern-day temperatures shooting upward) in the Climategate emails:

> *I tried hard to balance the needs of the science and the IPCC, which were not always the same.* I worried that you might think I gave the impression of *not supporting you well enough* while trying to report on the issues and uncertainties.[28] [emphasis added]

I can't imagine a traditional scientist trying to "balance the needs of the IPCC" (or any other organization) with his/her empirical scientific findings, but this is clearly okay in post-normal science to maintain the groupthink "consensus."

In the same vein, Briffa wrote to Mann (Sept. 22, 1999): "I know there is pressure to present *a nice tidy story* as regards 'apparent unprecedented warming in a thousand years or more'."

However, Briffa added: "In reality the situation is not quite so simple—I believe that *the recent warmth was probably matched about 1,000 years ago* [the Medieval Warm Period]." [emphasis added] [If this is so, by the way, then Mann's "hockey stick" with the jutting "blade" at the end is flat-out wrong.] In the end, Briffa abandoned the Medieval Warm Period to keep the alarmist "consensus" story "tidy."

I could hardly believe my eyes when I read these comments (and many others in the Climategate series): a *scientist* who would alter his/her findings to fit a preconceived, "nice tidy story." Wow.

But this apparent lapse in scientific standards becomes clearer when we realize that Briffa and Mann, and many (most?) other alarmists, are working within a "post-normal" scientific "consensus."

And when "consensus" is the top priority, as it is in groupthink, scientists must present a united front. Outliers and critics aren't allowed. In short, it is alarmist groupthink, not "science," that is leading us down the path of anti-fossil-fuel insanity.

Alarmist climate science avoids replication of results

Here's another more example of how "post-normal" science violates the standards of traditional science.

In traditional science, "reported results must be of such a nature that they can be independently replicated. Not until they are repeated under similar conditions by different and preferably skeptical investigators will they be finally accepted into the ranks of scientific knowledge."[29] In other words, scientific findings must be repeatable by other scientists.

Therefore, it's surprising (or perhaps not) that Climatic Research Unit head Phil Jones, when asked in 2004 to reveal the raw data for the temperatures given out by the CRU, replied: "We have 25 or so years invested in the work. Why should I make the data available to you, when your aim is to find something wrong with it?"[30] Eventually, in 2007, Jones was forced by the UK Information Commissioner's Office to release the data.

Similarly, when asked to produce for independent checking the data that led to the creation of the "hockey stick" graph, Michael Mann refused. Eventually the U.S. government stepped in to demand that Mann release his data, as required by U.S. law.[31]

Logic suggests to me that if I, as a scientist in any discipline, was forced to use non-falsifiable, "post-normal" science based on a subjective "consensus" of experts, rather than direct empirical and rigorously tested evidence, I'd be *especially* willing to hear and debate alternative theories; I'd be *especially* aware of the need to avoid claiming "certainty"; and I'd be *especially* eager to have my results confirmed or refuted by other scientists, even if those scientists were skeptical.

More: I'd aim to create a Red Team/Blue Team approach to my research and results: a group of fellow climate scientists, but a Loyal Opposition if you will, that *deliberately* seeks holes in my "consensus" argument.

If nothing else, I'd try to ensure that the IPCC reports include *minority* opinions, perhaps as appendices. It is quite simply *beyond belief* that the "2,000-plus scientists" who contribute to IPCC reports could have *full unanimity* on an issue as complicated as climate. So, what do the scientists who contribute to IPCC reports *disagree* on? Doesn't the public have a right to know this rather important information? You won't hear it from the IPCC or its supporters.

But then, I'd be assuming that alarmist climate scientists accept the traditional, skepticism-based approach to science; happily, the "post-normal" approach relieves them—at least in their own minds—of many of the professional and moral responsibilities of traditional scientists.

Following the scandal over Climategate, Mike Hulme did suggest: "[Climate] scientists need to be prepared to argue about their 'considered opinions', to embrace consensus but without closing down argument or suggesting that matters are settled."[32] Alas, Hulme's view was and still is a minority position among alarmist climate scientists, who much prefer to close down argument and suggest matters are "settled."

This fierce resistance to criticism (at least criticism from those outside the "consensus") is, I suppose, inevitable in "post-normal" science. Just as the solid ground of traditional science is empirical evidence, so the "solid ground" for "post-normal" science isn't solid at all—it's what most scientists in the discipline *agree upon* (in this case, the "consensus" that CO2 is the "control knob" of climate).

Having established a "consensus" among themselves, the next task of alarmist climate science is to get the politicians, media and public to accept their "subjective" theories as if they were the "hard" theories of traditional science.

They've done this—and succeeded brilliantly, for the most part—through the institutional prestige of the IPCC, but also by outright disinformation ("fake news"). Chapter 10 looks at some of the ways that alarmist climate science distorts the climate and environmental facts to fit alarmist theory.

Notes for Chapter 9

---

1 Climategate email from Phil Jones to Michael Mann dated July 8, 2004. Source: Climategate.

2 Climategate email from Tom Wigley to Timothy Carter dated April 24, 2003.

3 Richard Somerville. "A response to climate change denialism." Scripps Institute of Oceanography, Jan. 14, 2010.

4 I found this paraphrase of Popper in Steve Fuller, *Kuhn vs. Popper: The Struggle for the Soul of Science*. Toronto: Penguin Canada, 2006, p. 5. I haven't found the original quotation in Popper's works, although I have read most of his writings. However, the same idea in almost identical wording appears in A.C. Grayling's *Ideas That Matter: A Personal Guide for the 21st Century* (London: Phoenix Books, 2010, p. 206). Grayling writes of "Popper's belief that non-scientists can legitimately criticize science for failing to abide by its own avowed standards." Therefore, I am accepting this as a legitimate representation of Popper's views on the role of non-scientists.

5 "Monday was hottest day for global average temperature on record, as climate crisis bites." *The Guardian*, July 4, 2023. Available online. It's worth noting that *The Guardian* has abandoned all pretense of journalistic objectivity and is openly and proudly alarmist in its climate coverage.

6 I am indebted to retired engineer Ken Wilson for this insight. He discusses "post-normal" science in his PDF article "The IPCC vs. the Facts," available online.

[7] Wikipedia, "Post-normal science."

[8] IPCC, Climate Change 2001: Model Evaluation, "What is Meant by Evaluation?", Section 8.2.2.

[9] Mike Hulme, "The IPCC, consensus and science." Feb. 19, 2010. Mike Hulme website, https://mikehulme.org.

[10] Hulme, "IPCC, consensus and science."

[11] Stephen Schneider, *Laboratory Earth: The Planetary Gamble We Can't Afford to Lose.* New York: Basic Books, 1997, p. 67.

[12] Quoted in Patrick J. Michaels, *Sound and Fury: The Science and Politics of Global Warming.* Washington: Cato Institute, 1992, p. 83.

[13] IPCC AR6 (2023), Summary for Policymakers, "Possible Climate Futures," Section B.2.5, p. 16. Available online.

[14] Martin Goldstein & Inge Goldstein, *The Experience of Science: An Interdisciplinary Approach.* New York: Plenum Press, 1984, p. 4.

[15] E.D. Klemke, et al., *Introductory Readings in the Philosophy of Science.* Amherst, NY: Prometheus Books, 1998, p. 32.

[16] Karl Popper, *Unended Quest: An Intellectual Biography.* Glasgow: Fontana/Collins, 1976, p. 41.

[17] Economist Julian Simon once addressed a meeting of environmentalists and asked them the following questions: "How many people here believe that the Earth is increasingly polluted and that our natural resources are being exhausted?" Everyone agreed. He said, "Is there any evidence that could dissuade you?" Nothing. Again: "Is there any evidence I could give you—anything at all—that would lead you to reconsider these assumptions?" Not a stir. Simon then said, "Well, excuse me, I'm not dressed for church." You can find this story online with the search terms "Julian Simon" and "not dressed for church."

[18] Richard P. Feynman, *The Pleasure of Finding Things Out: The Best Short Works of Richard P. Feynman.* Cambridge, MA: Persius Books, 1999, p. 111. Feynman won the Nobel Prize for physics in 1965.

[19] For example, Charles Onians, "Snowfalls are now a thing of the past." *The Independent,* March 20, 2000. This article is one of a list of alarmist prediction failures presented by Steven J. Milloy & Myron Ebell in "Wrong Again: 50 Years of Failed Apocalyptic Predictions." *Competitive Enterprise Institute,* Sept. 9, 2019. Available online. See also David Siegel, "The Master Shortfall List of Failed Climate Predictions," *Shortfall,* April 27, 2022. Available online. Both websites make for hilarious reading.

[20] Popper, *Unended Quest,* pp. 24,38.

[21] Popper, *Conjectures and Refutations: The Growth of Scientific Knowledge.* New York: Routledge & Kegan Paul, 2007 (1963), p. 48.

[22] Steve Fuller, *Kuhn vs. Popper: The Struggle for the Soul of Science.* Cambridge: Icon Books, 2006 (2003), p. 222.

[23] Popper, *The Open Society and Its Enemies,* Vol. 1. Princeton, NJ: Princeton Univ. Press, 1971 (1943), p. 4.

[24] Einstein, "What is the Theory of Relativity?" *Ideas and Opinions.* New York: Laurel Edition, 1954, p. 227. This article was originally a letter to the London *Times.* Available online.

[25] Charles Darwin, "Introduction," *The Origin of Species.* New York: Collier Books, 1962 (1859), p. 26.

[26] Irvin L. Janis, *Groupthink: Psychological Studies of Policy Decisions and Fiascoes.* Boston: Houghton Mifflin, 1982, p. 3.

[27] *Groupthink*, p. vii.

[28] Keith Briffa, Climategate email to Michael Mann, April 29, 2007.

[29] Victor J. Stenger, *God: The Failed Hypothesis: How Science Shows that God Does Not Exist* (New York: Prometheus Books, 2008, p. 25. This book has a listing of the requirements of good (traditional) science, almost none of them met by alarmist climate science.

[30] Andrew Orlowski, "Global Warming ate my data." *The Register*, Aug. 13, 2009. In 2009, the CRU revealed that it had destroyed the original data upon which much of the world temperature record is based, making it impossible for a third party to reconstruct the record.

[31] For details on Stephen McIntyre's attempt to replicate Mann's work see McIntyre and Ross McKitrick, "The IPCC, the 'Hockey Stick' Curve, and the Illusion of Experience." Washington Roundtable on Science & Public Policy, George C. Marshall Institute, Nov. 18, 2003. Available at www.marshall.org/pdf/materials/188.pdf.

[32] Hulme, "The IPCC, consensus and science." Cited above.

Chapter 10: The dishonesty of alarmist climate science

"Well, now that we have seen each other," said the Unicorn, "if you'll believe in me, I'll believe in you. Is that a bargain?"

As noted in Chapter 9, alarmist climate science operates on a "post-normal" model, based not on rigorous empirical testing of theories (since "we can't put Earth in a test tube" and we can't have "facts" about the future) but the informed (i.e., subjective) opinions of "experts," scientists who have studied the planet's weather and climate history and then created computer-model-based theories to explore what the future climate might, maybe have in store if we keep adding carbon dioxide.

So far so good. Experts obviously know more than the average person about their subject.

But these climate "experts" then took a further step: they pronounced their subjective "consensus" theories of climate catastrophe as "The Science"—and declared that The Science was not to be challenged by non-scientists, scientists from other disciplines, or even non-alarmist climate scientists, in part to ensure that the alarmists' beautiful theories could not be slain by any "ugly fact," as Huxley put it.

As additional protection, alarmist climate scientists also undertook or supported a campaign of disinformation for the public, media and politicians—"fake news"—as you'll see in this chapter.

But before we get into climate-science dishonesty, let's review what the gold standard for a scientist—traditional and also, I'd hope, "post-normal"—ought to be in relation to non-scientists. Physicist Richard Feynman was very clear that scientists must be scrupulously honest with lay people (and "lay people" includes the media and politicians):

> *You should not fool the layman when you're talking as a scientist.* ... I'm talking about a specific, extra type of integrity that is ... *bending over backwards to show how you are maybe wrong*, that you ought to have when acting as a scientist. And this is our responsibility as scientists, certainly to other scientists, and I think to laymen.[1] [emphasis added]

Similarly, James Lovelock, creator of the "Gaia" hypothesis and himself an ultra-alarmist for a time before taking a more skeptical and "realist" position (see **Sidebar 6: James Lovelock** at the end of this chapter), also makes the sacredness of scientific truth (presumably including "post-normal" science) quite clear when he observes:

> *Fudging the data in any way whatsoever is quite literally a sin against the holy ghost of science.* I'm not religious, but I put it that way because I feel so strongly. It's the one thing you do not ever do. You've got to have standards.[2] [emphasis added]

"You've got to have standards." Let's see how well one of the most prominent climate alarmists, the late Stephen Schneider, approached this question of scientific standards, of not fudging the data, and of not "fooling" lay people.

Exaggeration and 'scary scenarios'

In an interview with *Discover* magazine in 1989 Schneider stated quite clearly that alarmist climate scientists may have to *exaggerate* the problem to get the public's attention and acquiescence. He frequently complained that portions of his comment were taken out of context to make him look bad, so here is his statement in its entirety:

> On the one hand, as scientists we are ethically bound to the scientific method, in effect promising to tell the truth, the whole truth, and nothing but—which means that we must include all the doubts, the caveats, the ifs, ands, and buts. On the other hand, we are not just scientists but human beings as well. And like most people we'd like to see the world a better place, which in this context translates into our working to reduce the risk of potentially disastrous climatic change.

> To do that we need to get some broad-based support, to capture the public's imagination. That, of course, entails getting loads of media coverage. *So we have to offer up scary scenarios, make simplified, dramatic statements, and make little mention of any doubts we might have.*
>
> This "double ethical bind" we frequently find ourselves in cannot be solved by any formula. Each of us has to decide what the right balance is between being effective and being honest. I hope that means being both.[3] [emphasis added]

Schneider was, I have no doubt, a man of integrity who in his personal life highly valued honesty and so, paradoxically, he is honest (an honest liar!) in telling us that he is sometimes dishonest ("scary scenarios," "little mention of doubts") in presenting his alarmist climate views.

He even admitted in another interview that maybe The Science isn't quite as "settled" and "certain" as he and other alarmists would like the media, politicians and public to believe. On Australian radio he spoke of his "lingering frustration" that, in trying to anticipate the future,

> ... *we end up with a maddening degree of uncertainty.* We end up with scenarios which, if we're lucky, give us mild outcomes and we end up with scenarios that, if we're unlucky, give us catastrophic outcomes. ... *We fight amongst ourselves bitterly about the relative likelihoods of these, and have virtually no agreement....*[4] [emphasis added]

Elsewhere, in a *Scientific American* article critical of Bjorn Lomborg's *The Skeptical Environmentalist: Measuring the Real State of the World,* he attacks Lomborg for daring to assume the "real state of the world" is known (Schneider apparently didn't see the irony that he and his alarmist colleagues are very big on claiming, at least to the public, that they know "the real state of the climate"):

> I readily confess a lingering frustration: Uncertainties so infuse the issue of climate change that it is still impossible to rule out either mild or catastrophic outcomes let alone provide confident probabilities for all the claims and

counterclaims made about environmental problems. ... Uncertainties are so endemic in these complex problems that suffer from missing data, incomplete theory and nonlinear interactions.[5] [emphasis added]

How very curious, to quote Alice. That's not what we—the public, politicians and the media—are told.

We're told The Science is settled and beyond question, as far from "maddening uncertainty" as you can get, to the point where there is, as climate modeler Andrew Weaver put it, "no debate" on this issue within climate circles because "making the public believe that such a debate exists is precisely the goal of the denial industry. ... *Scientific debate over global warming would therefore imply uncertainty.*[6] [emphasis added]

It's hard to reconcile Weaver's view of climate science (no uncertainty) with Schneider's in his more revealing moments (maddening uncertainties), but why would Schneider lie about this when, by telling the truth, he's actually undermining his own alarmist case?

Climate policies based on 'maddening uncertainty'

What this means is that political leaders are asked to make decisions that could radically change the world by damaging or destroying our fossil-fuel-based economic system—a system that has produced and is producing increasing prosperity for billions—based on this so-called "settled" but actually "maddeningly uncertain" alarmist "Science."

Schneider's comments, which are on public record, show that, contrary to Feynman's maxim, we, lay people, are clearly being deliberately "fooled," are we not? And not just by Schneider, as I've tried to show.

And are alarmist climate scientists, as Feynman recommends, "bending over backwards" to show us where their science "may be wrong," making us aware of the possibility that alarmist climate science predictions of disaster may not be a "certain" as we've been told?

Absolutely not, at least not in public, by the vast majority of alarmist climate scientists (Schneider's honesty is rare). Instead, those who dare to criticize alarmist climate science are called "deniers" and demonized. A very strange science, this "post-normal" science, which is more reminiscent of the Spanish Inquisition than the rational, open, scrupulously honest inquiry we hope for from science ("you've got to have standards").

And, just to repeat: Despite his cheerleading for extreme climate alarmism, and as we saw in Chapter 8, Schneider didn't actually think we were facing a planetary climate "catastrophe," writing: "In the sense that it [climate change] will be an urgent catastrophe which will eliminate all life, *that kind of statement is absolutely unwarranted.*"[7]

Good to know. Too bad this rather important bit of information hasn't gotten to the Extinction Rebellion crowd and the millions of young people who've been terrified into believing, like Greta Thunberg, that "climate change" has "stolen" their futures.

World not overwhelmed by 'extreme' climate events

Not having to worry about the inconvenient rules of "normal" science gives climate alarmists more free time to produce an enormous amount of material aimed at showing us, the public, that thanks to "climate change" our world is facing environmental and human collapse from increased floods, storms, hurricanes, forest fires, sea-level rise, heat waves, industrial pollution, a "sixth mass extinction," and many more threats.

Almost all of these claims are "fake news," based on alarmist propaganda rather than facts, as I'll show you in more detail below, but let me give you one example from a recent news story:

B.C. Premier David Eby said he experienced "the devastating threat of climate change" in 2020 while staying at a family cabin in B.C.'s wildfire zone. "There was a forest fire across the water and at night we could literally watch the trees burn and explode, literally explode, because of the heat from the fire," he said.[8]

In contrast, Figure 10.1 shows data from the Canadian Interagency Forest Fire Centre on the annual number of fires in Canada from 1990-2021.[9]

Curiously, the average number of forest fires per year in Canada is *down* over the past 30 years, not up as Eby and many others have been led to believe by alarmists (yes, 2023 has seen a high number of fires, many of them human-caused, but not as many as, say, 1995, 2006, or 2012, among other years, which rather spoils the alarmist end-of-the-world-in-2023 forest-fire narrative).

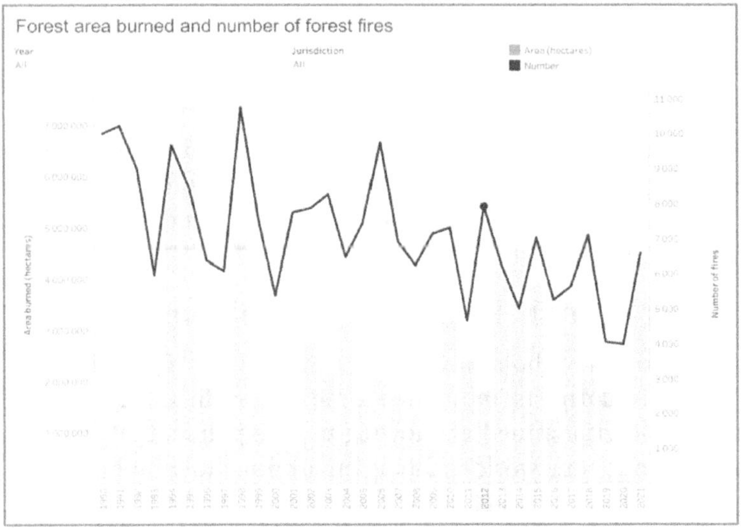

Figure 10.1: Number of fires in Canada from 1990-2021 is down, not up. Source: National Forestry Database

It's odd, isn't it, that there would be *fewer* forest fires in Canada in the past two decades if the climate was warming up as much as the alarmists tell us; fewer fires is not so odd if "global warming" had stalled for those decades, which it has, as we saw in Chapter 4.

Undersea volcano caused most of 'record' warming in 2023

Canadians are being told that 2023's busy summer fire season is due to "record" "global warming"; in fact, the increase in heat and fires is almost certainly due to the effect of the large undersea volcanic explosion near Tonga (noted in Chapter 5) and a large impending El Niño. But, hey, why spoil a good alarmist story with facts?

Similarly, the acreage burned by wildfires in the United States is also way down over the 20th century, as Figure 10.2 shows, which is odd if "global warming" was an increasing menace as the century proceeded. Shouldn't acreage burned increase as "global warming" increased?

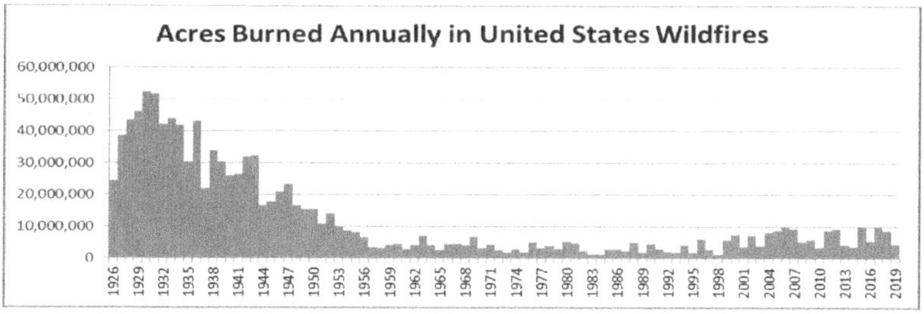

Figure 10.2: Forest fire areas have dropped considerably in the United States. Source: Wikipedia, "Total Wildland Acres Burned Annual (1926-2019) in the United States"

Now, statistically, alarmists can make the forest-fire situation look more dire by focusing on economic damage—recent fires do more *economic* damage today.

But that's only because the Earth is more populous; the increased economic damage isn't because there are more fires but because more people are living in fire-prone areas.

(And the same is true for increases in flood damage; the number and intensity of floods is about the same, or even less, but more people are living in flood-prone areas than ever before.)

Even so, *fewer* numbers of fires for Canada and the U.S. are hard to square with the idea that the planet has been and is warming so drastically that the number of forest fires is going out of control or that we are, as a species, existentially threatened by them.

'Extreme weather' more common in the past

One reason forest fires aren't the huge problem alarmists would have us believe: we have better fire-suppression measures today and better methods of reacting to fires to save lives.

If this is so, the logical conclusion is that *even if* global warming makes forest fires (or heat waves or floods or droughts) more numerous and more intense, humanity is quite capable of dealing with them. But in the meantime, if we look at the actual data, the alarmist narrative of more fires than in the past is clearly false.

Also false are almost all of the other alarmist prophecies of "extreme weather" doom. A study by M.J. Kelly published in 2016 examined extreme weather data since 1900. The study's conclusion:

> It is widely promulgated and believed that human-caused global warming comes with increases in both the intensity and frequency of extreme weather events. A survey of official weather sites and the scientific literature provides strong evidence that the first half of the 20th century had more extreme weather than the second half, when anthropogenic global warming is claimed to have been mainly responsible for observed climate change.[10] [emphasis added]

Wow! Our parents and grandparents somehow lived through far more "extreme weather" than we are currently experiencing. Who knew?

Kelly even writes that he is puzzled at how the claims of "extreme weather" could be so far removed from reality:

> *The disconnect between real-world historical data on the 100-years' time scale and the current predictions* provides a real conundrum when any engineer tries to make a professional assessment of the real future value of any infrastructure project which aims to mitigate or adapt to climate change. What is the appropriate basis on which to make judgments *when theory and data are in such disagreement?*[11] [emphasis added]

(We already know the answer to this question: alarmist climate science goes with the theory, not the data.)

The study quoted above includes a graph (Figure 10.3) showing that you were far more likely to die from an "extreme weather" event before 1950 than today! People in the 1920s, '30s and '40s faced some major climate challenges, including the prolonged drought of the Dirty Thirties. But did our parents and grandparents live in *existential* terror of "extreme weather"? I doubt it, so why are we so fearful of our lives in current, actually *safer* times?

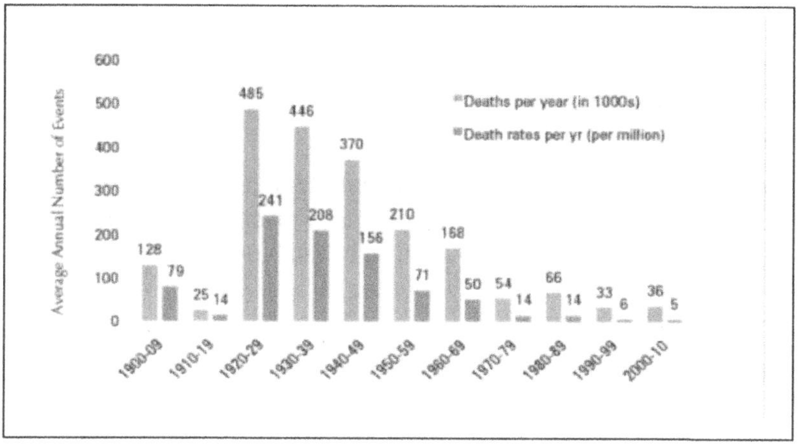

Figure 10.3: Extreme weather was far more common, and dangerous, in the first half of the 20th century

A more recent study (2021) comes to the same conclusion: that there is no clear trend of increased "extreme weather" events, nor are we at increased risk of being a victim of "extreme weather":

> The analysis [in the study] is then extended to some global response indicators of extreme meteorological events, namely natural disasters, floods, droughts, ecosystem productivity and yields of the four main crops (maize, rice, soybean and wheat). *None of these response indicators show a clear positive trend of extreme events.*

> In conclusion on the basis of observational data, *the climate crisis that, according to many sources, we are experiencing today, is not evident yet.*[12] [emphasis added]

David Attenborough: False prophet of doom

Many people get their alarmist view of "climate change" from nature documentaries, including those featuring David Attenborough, the 90-plus-year-old British presenter of countless films on the natural world, most with the BBC, and an ardent proselytizer on the perils of global warming.

For example, a 2019 Attenborough documentary for BBC, *Climate Change: The Facts,* opens with a montage of climate and industrial horrors: floods, hurricanes, fires, gas flares, smokestacks, suffering animals and people.

Anyone seeing just this film would conclude that our planet is a truly horrible place to live. And, indeed, many people, including many young people and even children, actually believe they are going to die due to "climate change" (we're looking at you, Greta Thunberg and Extinction Rebellion).

This just means they would rather believe media reports than the evidence of their own senses, since most people in the developed world, at least, have perfectly good lives (this includes Greta and her ER brethren who have somehow convinced themselves that they live on the edge of "extinction").

We don't live on a blasted planet; far from it, which is obvious if we're willing to accept facts rather than alarmist propaganda. Those who believe otherwise are victims of the enormous psychological damage caused by alarmist disinformation.

For example, watching *Climate Change: The Facts*, one would never guess that deaths from "extreme" natural events like hurricanes and floods are much, much less prevalent than was the case even a century ago. Figure 10.4 is a chart from *Our World in Data* showing annual deaths from natural disasters:

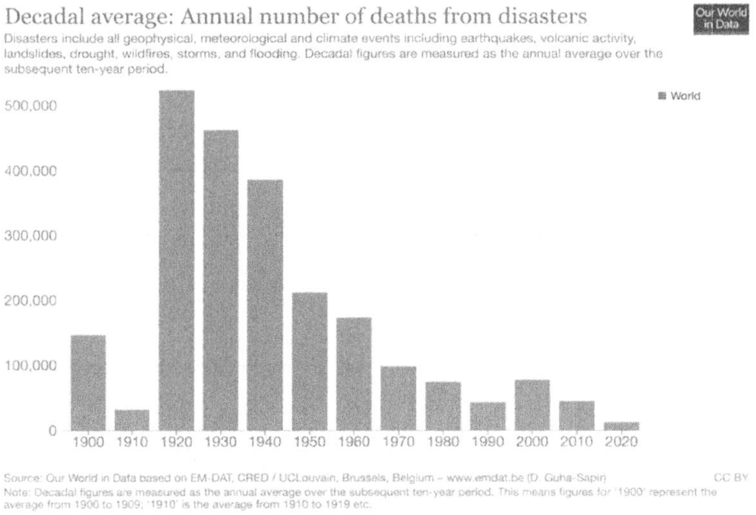

Figure 10.4: Average number of deaths from disasters over the past 120 years—way down! Source: Our World in Data

Does this look like a world where millions of people are currently in imminent peril to their lives from "climate change," as the Attenborough documentary wants us to believe?

On the contrary, a search of the *Our World in Data* website offers ample evidence that in terms of vulnerability to "extreme" weather events, humanity has never been better protected, thanks in part to the fossil-fuel-based, advanced technology that the documentary deplores.

Nations that don't have this advanced technology are, naturally, harder-hit when nature strikes, but surely the solution isn't to deprive poorer countries of fossil-fuel energy, as the documentary urges, thereby making them even *more* vulnerable to nature's whims.

China and India have no intention of giving up the advantages of fossil fuels as they grow their economies; why should other developing nations be hamstrung as they try to improve the lives of their people?

It's also worth noting that "extreme" weather events are not unusual in human history. As noted above, these events do more damage today, in dollar and human terms, only because, with a planetary population of eight billion, more people are living in areas likely to be affected by these events (e.g., the New Orleans flood) and, thanks to a higher standard of living, have more assets to be destroyed.

In British Columbia, the Sumas Prairie flood of November 2021 was disastrous, causing more than $2-billion in damage, killing more than 600,000 farm animals, making thousands homeless and killing five people.

But this flood was not "unprecedented" or even unanticipated—major flooding in this area also occurred in 1894, 1948, 1972, and 2007.[13] Were all these floods due to "climate change" (1894? 1948? 1972?)? Or were they just what happens when you drain a lake so that, inevitably, the lake will try to return?

The "solution" to "extreme" weather events like Sumas Prairie or the New Orleans disaster isn't to cut off fossil fuels so the planet might warm a tiny bit less. The solution is to ensure that flood protections such as dikes and embankments are in place, regularly checked, and repaired if necessary.

It's on public record that this (yes, expensive) upkeep was not performed either in New Orleans or B.C.,[14] so do we blame "climate change" for these disasters? Or ourselves and our governments for not preparing for the strong possibility of flooding in certain areas?

Climate 'disasters' not increasing

That said, watching *Climate Change: The Facts,* you would never know that the number and intensity of "weather disasters" is *not increasing*, as noted above.

Figure 10.5, below, shows the number of global weather and climate disasters from 2000 to 2022—the average number is actually *down* a bit![15] You'd think the number of "extreme" weather events would be noticeably increasing thanks to out-of-control "global warming"....

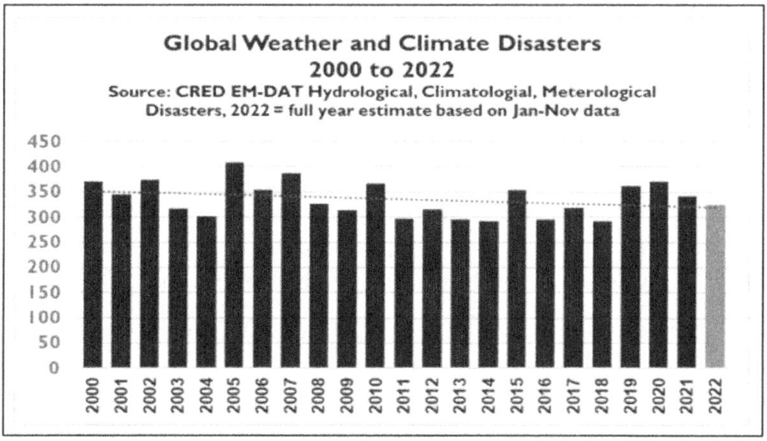

Figure 10.5: Number of global weather and climate disasters is falling, not rising. Source: Center for Research on the Epidemiology of Disasters (CRED)

Nor, watching *Climate Change: The Facts,* could one guess that fossil fuels, far from ruining people's lives, are bringing most of the world's people out of extreme poverty by providing relatively cheap energy—energy that is less polluting and damaging to human health than, say, burning wood or charcoal.

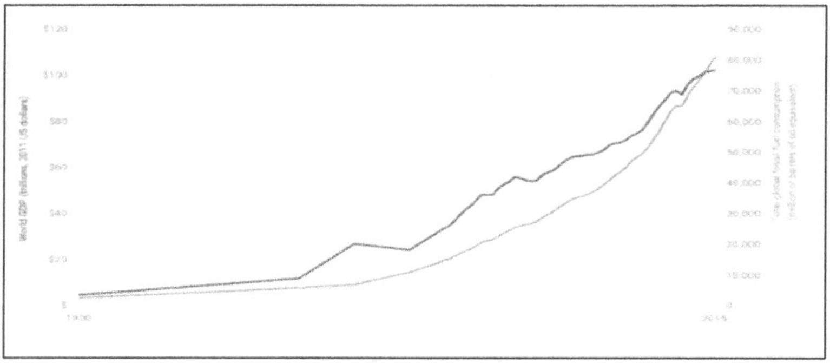

Figure 10.6: Correlation of GDP and fossil-fuel use. Source: Brookings Institute, using data from Our World in Data

Figure 10.6 shows the relationship between the use of fossil fuels and global GDP, which is a reasonably good measure of overall human well-being. The top line is GDP; the lower line is total global fossil-fuel consumption.

The correlation is striking: fossil fuels (i.e., modern industry, technology and agriculture) have radically *enhanced* human life for most people on the planet, a fact that *Climate Change: The Facts* distorts and denies. Yes, we face climate challenges now and in the future, but to say the planet is somehow being destroyed by these challenges is to put ideology over observation and just common sense.

The chart below (Figure 10.7), also from *Our World in Data*, shows the fall in poverty and extreme poverty from 1820 to 2015. The top line shows the fall in the numbers of those living in poverty (defined as less than US$2.15 a day); the lower line shows the fall in numbers of those living in *extreme* poverty (less than US$1/day).

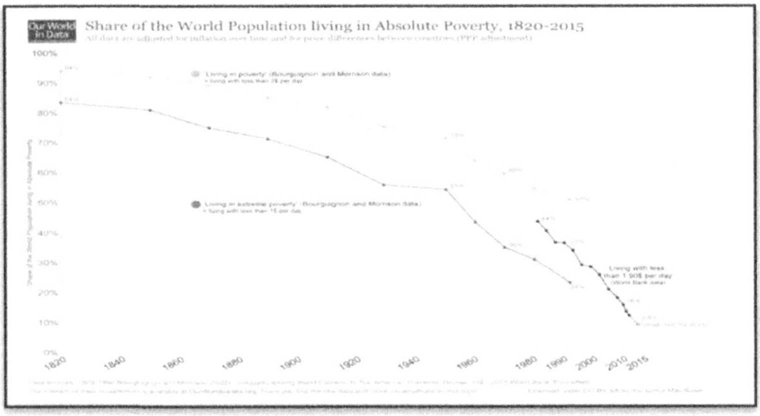

Figure 10.7: Fall in poverty and extreme poverty 1820-2015. Most of this improvement is due to fossil-fuel energy. Source: Our World in Data

This improvement in living conditions is almost entirely due to the rise of industry, technology, and modern agriculture using (relatively) cheap fossil-fuel energy and artificial fertilizers, both unmitigated evils from the point of view of the Attenborough documentary.

That the Attenborough documentary ends with a (now-familiar) diatribe from Greta Thunberg should be a big clue that the film's "facts" are directed more at the audience's emotions than its reason, with the goal of creating fear rather than offering a balanced assessment of a complex issue.

It's as if a BBC camera crew went to Vancouver's Downtown Eastside, filmed the poverty and disorder there for a few days, and produced a documentary called *Vancouver: The Facts.* Yes, the "facts" about the Downtown Eastside would be true as far as they go—there is much poverty and disorder there—but the BBC film would be offering a wildly distorted picture of the larger truth that Vancouver is a very livable city for most of its citizens.

Similarly, Attenborough's "facts" are true, as far as they go, but are cherry-picked for shock value. A more balanced perspective reveals that the planet is *not* getting worse for human beings, despite "climate change," and is actually getting better—when we aren't wrecking our economies in a futile war against "carbon emissions."

My point is: if many or most members of the public believe we are facing a climate disaster, it might be because they watched an Attenborough documentary, Al Gore's *An Inconvenient Truth,* or other media that distort the facts (and even make up "facts"[16]) to give the public that apocalyptic impression. These distortions are part of the "post-normal," "woke" approach to science: if you *feel* climate change is a threat, then it is. Reality is optional.

Some scientists 'embarrassed' by climate doom-saying

Another reason why the public gets a distorted picture of climate issues is mentioned by Steven Koonin, a physicist and former U.S. undersecretary for science, in his 2021 book *Unsettled: What Climate Science Tells Us, What It Doesn't, and Why It Matters.*

He writes that information in the IPCC's reports is often exaggerated for the public by the writers of the IPCC Summary for Policymakers, by alarmist environmental groups (including the United Nations itself), by politicians, and by the media (we'll see more on this exaggeration in Chapter 11).

He writes of a meeting of the American Physical Society he attended that was convened to "stress test" the state of climate science. His observation, based on these discussions:

> Government and UN press releases and summaries *do not accurately reflect the reports themselves.* ... Distinguished climate experts (including report authors themselves) *are embarrassed by some media portrayals of the science.* This was somewhat shocking.[17] [emphasis added]

It *is* shocking, or should be, to anyone who cares about truth. Koonin also concluded that "the science is insufficient to make useful projections about how the climate will change over the coming decades, much less what effect our actions will have on it." Which is very much what this book is arguing.

In other words, politicians, environmentalists and the media tend to magnify the IPCC's claims. Politicians and environmentalists exaggerate for ideological political reasons, the media because "if it bleeds it leads" (and sells papers, or their digital equivalent). "World not warming that much" or "Disasters not increasing" doesn't make a striking headline.

'Adjusting' the climate data

Climate alarmists have another trick up their sleeve: they can (and do) "adjust" or "homogenize" previous temperature data to make our time seem warmer than earlier times that actually *were* warmer.

For example, the 1930s were at least as hot, and likely hotter, than today. This creates a problem for alarmists who want the public to believe today's warming is "unprecedented." But there's a simple fix: change the earlier data!

For example, Figure 10.8, next page, originally from the Goddard Institute for Space Studies in 1999, shows the temperature averages in the United States from 1880 to 2000. Note that the peak of the 1930s (circled at top) in Figure 10.8 was hotter than 2000 (circled at right), which was not a good look for climate alarmists.

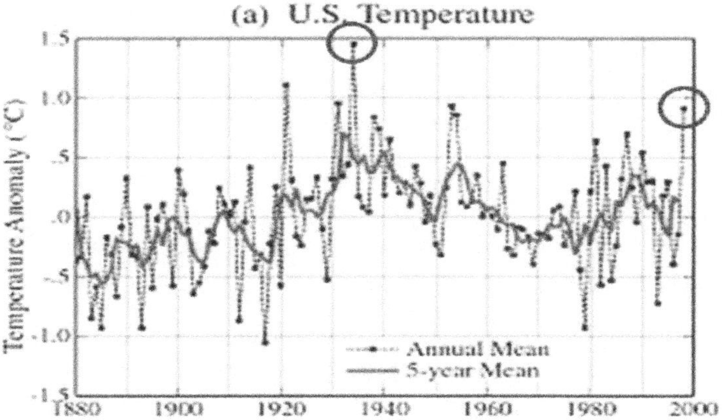

Figure 10.8: GISS temperature chart for U.S. shows a very warm 1930s and cooler 2000. Source: Steven Goddard

Fortunately, a solution was at hand! Just "adjust" the temperature record to *reduce* 1930s temperatures and *raise* the 2000 temperatures (Figure 10.9). Voila! 2000 shows "record" warmth again![18] Alarmist theory saved!

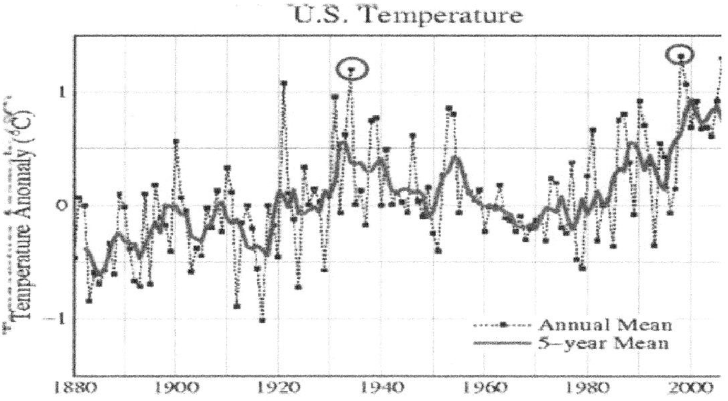

Figure 10.9: GISS temperatures "adjusted" to reduce 1930s warming and make 2000 look warmer

The IPCC has done the same thing with the Medieval Warm Period and, indeed, all periods that were warmer than today in the past 10,000 years, such as the Roman warming (2,000 years ago), the Minoan warming (3,500 years ago) and the Holocene Optimum (8,000 years ago)—they have been "adjusted" out of existence so the IPCC can claim today is the warmest in 125,000 years.[19]

To sum up, the mass media, along with alarmist climate scientists and environmentalists, are distorting and exaggerating the IPCC's somewhat less-alarming findings to the point where even alarmist climate scientists occasionally find themselves cringing.

This means we, as citizens, can't trust alarmist climate scientists to level with us; we have to do the research ourselves—the facts are out there, but we have to dig for them. The alarmists certainly won't help us and in fact will do everything they can to hinder us.

So when you are told that the environment is going to hell, that "extreme weather events" are threatening us and our planet, that the planet is experiencing "record temperatures," that sea levels are rising uncontrollably, that we're being increasingly scorched by forest fires and drowned by floods—in short, we're all gonna die if we don't do what the alarmists want—I'd suggest approaching these claims with a ton of skepticism and preferably some background research of your own. Almost all of these claims are misleading, exaggerated, or just plain wrong.

As astute readers may have guessed, I am not a big fan of the Intergovernmental Panel on Climate Change. In Chapter 11 I'll offer several reasons why we should also regard the IPCC's reports with extreme caution.

Notes for Chapter 10

---

[1] Richard P. Feynman, "Cargo cult science: The 1974 Caltech Commencement Address," in *The Pleasure of Finding Things Out: The Best Short Works of Richard P. Feynman*. Cambridge, MA: Persius Books, 1999, p. 212.

[2] Leo Hickman, "Humans are too stupid to prevent climate change." Interview with James Lovelock. *The Guardian*, March 29, 2010.

[3] Schneider is quoted in Jonathan Schell, "Our Fragile Earth." *Discover*, October, 1989, pp. 45-48. See also Stephen Schneider, "Don't Bet All Environmental Changes Will Be Beneficial," *APS* (American Physical Society) News, August/September 1996, p. 5, where he defends himself against what he considers misrepresentation of his views.

[4] Earthbeat, "Skeptical Environmentalist Debates Critics." *Australian Broadcasting Corp.,* Oct. 10, 2001.

[5] Stephen Schneider, "Global Warming: Neglecting the Complexities." *Scientific American,* January, 2002. For (many) more examples of alarmist climate science attempts to deceive the public and decision-makers, see my book *False Alarm*, Chapter 7, "Open Science and its Enemies: The Global Warming Paradigm."

[6] Andrew Weaver, *Keeping Our Cool: Canada in a Warming World.* Toronto: Viking Canada, 2008, pp. 22-23.

[7] Stephen Schneider, *Science as a Contact Sport: Inside the Battle to Save Earth's Climate.* Washington, DC: National Geographic, 2009, p. 94. The fuller quotation is as follows: "I do not believe that [environmental] change will necessarily be catastrophic, although the issue of catastrophe is almost irrelevant. ... It is like many other pollution issues. I have a sense of urgency in the view that we need to consider now those actions which increase our options. In that sense, it is urgent. In the sense it will be an urgent catastrophe that will eliminate all life, that kind of statement is absolutely unwarranted."

[8] Dirk Meissner, "B.C. adds $180M to help communities to prepare for natural disasters." *Times Colonist*, Feb. 22, 2023.

[9] Canadian Interagency Forest Fire Centre, graph at https://ciffc.ca/fire-information/wildfire-graphs. See also Marc Montgomery, "Study says Canadian forest fires have declined since 1989 peak." *Radio Canada International,* Nov. 19, 2020.

[10] M.J. Kelly, "Trends in Extreme Weather Events since 1900 – An Enduring Conundrum for Wise Policy Advice." *Journal of Geography and Natural Disasters*, Feb. 17, 2016. Available online in pdf format.

[11] Kelly, *Ibid.*

[12] Gianluca Alimonti, et al., "A critical assessment of extreme events trends in times of global warming." *European Physical Journal Plus*, Nov. 29. 2021.

[13] Wikipedia, "Sumas Prairie."

[14] Brenna Owen, "BC aware of dike problems before destructive flooding in 2012: documents." Victoria *Times Colonist,* Sept. 14, 2023.

[15] Graph found in Roger Pielke, Jr., "Has the IPCC outlived its usefulness?" *Honest Broker* blog, March 23, 2023. The CRED disaster database can be found at https://www.emdat.be/.

[16] A recent *Netflix* film on the Arctic walrus, *Our Planet,* narrated by Attenborough, was deliberately edited to claim that walruses sometimes hurl themselves off cliffs due to lack of space on the sea ice caused by "global warming." In reality, as the filmmakers knew at the time but chose to ignore, the walruses were stampeded by a marauding polar bear and these cliff-falls are not uncommon. For details, see Susan J. Crockford, "New footage reveals Netflix faked walrus climate deaths." *Polar Bear Science,* Nov. 19, 2020. Available online.

[17] Steven Koonin, *Unsettled: What Climate Science Tells Us, What It Doesn't, and Why It Matters.* Dallas: Ben Bella Books, 2021, p 4.

[18] Steven Goddard, "Uncorrupted U.S. temperature data showed cooling from 1930 to 1999." *Real Climate,* June 11, 2012. Available online at https://stevengoddard.wordpress.com/2012/06/11/uncorrupted-us-temperature-data-showed-cooling-from-1930-to-1999. On the website the author has created a gif comparing the two graphs to highlight the "adjustment."

[19] IPCC, Sixth Report, Working Group I, Summary for Policymakers, p. 6.

Sidebar 5: James Lovelock: From ultra-alarmist to realist

James Lovelock (1919-2022), a medical doctor and ecologist, was an ultra-climate-alarmist who believed human beings were a "fever" that was "perceptibly disabling the planet like a disease."[1]

As a result of this disease (us), he asserted: "Before this century [the 21st century] is over billions of us will die and the few breeding pairs of people that survive will be in the Arctic where the climate remains tolerable."[2] For the ultra-alarmist Lovelock, what was most important "is the health of the Earth, not the health of people"[3]—a disturbingly anti-humanist position.

But over time Lovelock's views evolved into a much more *realistic* position. And he wasn't afraid to admit that he might have been wrong in his earlier extreme views, in part because he valued scientific truth over alarmist ideology.

And, looking at the alarmist climate data, Lovelock observed a lot of "fudging" (as he called it) of the truth:

> The great climate science centres around the world are more than well aware how weak their science is. If you talk to them privately they're scared stiff of the fact that they don't really know what the clouds and the aerosols are doing. They [clouds and aerosols] could be absolutely running the show. We haven't got the physics worked out yet.
>
> *We do need skepticism about the predictions about what will happen to the climate in 50 years*, or whatever. It's almost naive, scientifically speaking, to think we can give relatively accurate predictions for future climate. *There are so many unknowns that it's wrong to do it.*[4] [emphasis added]

Support for nuclear power

He also became more realistic about the claims that renewable energy sources—including solar and wind—could somehow take us to the next, carbon-neutral level of civilization, as claimed by so many environmentalists and green politicians.

Instead, Lovelock outraged many Greens by endorsing nuclear power to reduce carbon emissions and by criticizing as irrational the green movement's fear of nuclear energy.[5] He wrote:

> Renewable energy sounds good, but so far it is inefficient and expensive. It has a future, but we have no time now to experiment with visionary energy sources: *civilization is in imminent danger and has to use nuclear energy now, or suffer the pain soon to be inflicted by our outraged planet.*
>
> ...
>
> Nuclear energy is merely the medicine that sustains a steady secure source of electricity to keep the lights of civilization burning until clean and everlasting fusion, the energy that empowers the sun, and renewable energy are available.[6] [emphasis added]

Lovelock cited reams of statistics showing that nuclear plants, far from being risky, are in fact far safer than any other type of energy generation.[7]

This more realistic Lovelock even had good things to say about climate skeptics:

> I think you have to accept that the skeptics have kept us sane—some of them, anyway. They have been a breath of fresh air. *They have kept us from regarding the science of climate change as a religion. It had gone too far that way.* There is a role for skeptics in science. They shouldn't be brushed aside. It is clear that the angel side wasn't without sin.[8] [emphasis added]

Lovelock is a fine example of a scientist who moved from an alarmist (even ultra-alarmist) position to a much more realistic view based not on ideology but on scientific facts: Lovelock was honest enough to recognize that the *facts* supported a realist position, not alarmist dogma, when it came to coping with climate change.

It's a pity more so-called climate "scientists" haven't taken a similar hard, cold look at the data and joined Lovelock on the realistic side of the mirror.

Notes to Sidebar 5: James Lovelock

---

[1] James E. Lovelock, *The Revenge of Gaia*, Toronto: Penguin Books, 2006, pp. 55-57.

[2] Lovelock, "The Earth is about to catch a morbid fever." *The Independent*, Jan. 20, 2006.

[3] *Revenge*, p. 8.

[4] Leo Hickman, "Humans are too stupid to prevent climate change." Interview with James Lovelock. *The Guardian*, March 29, 2010.

[5] *Revenge*, p. 112.

[6] *Revenge*, p. 14.

[7] *Revenge*, p. 131.

[8] Charles Clover, "Grandaddy of green, James Lovelock, warms to eco-skeptics." *Sunday Times*, March 14, 2010.

Chapter 11: The IPCC's 'government-approved' science

If one drinks much from a bottle marked 'poison,' it's almost certain to disagree with one sooner or later.

Ground zero for Net Zero is the Intergovernmental Panel on Climate Change (IPCC), which was established in 1988 "to provide policymakers with regular scientific assessments on the current state of knowledge about climate change."[1]

So far, so good. Given that in 1988 scientists were aware that CO2 levels were increasing at more than the "natural" pace, and that CO2 was a factor in climate warming, it made sense to keep an eye on this development. But then the IPCC went immediately off the rails.

Logic and the tenets of good science suggest this assessment of climate change would include *all* the possible elements of climate change, natural and human.

In fact, the IPCC's mission statement was much narrower: it was not to explore the many factors involved in creating climate change but

> to assess on a *comprehensive, objective, open and transparent basis* the scientific, technical and socio-economic information relevant to understanding the scientific basis of risk of *human-induced climate change*, its potential impacts, and options for adaptation and mitigation.[2] [emphasis added]

In short, the IPCC was biased from the start: when it comes to "climate change," *only human-caused factors need apply*. Since climate change has been going on for billions of years entirely due to natural causes, one would think that the human component would be a relatively minor add-on, worthy of some study, for sure, but just one of dozens of climate-change factors.

Not so for the IPCC and, having set out its mission, it proceeded (after a couple of false starts, discussed below) to—surprise!—find what it and its alarmist promoters were looking for: evil humans were strongly affecting the climate! Another term for the IPCC's process is "confirmation bias"—selection of data to fit a preconceived theory.

'Review by governments' destroys IPCC credibility

As if this wasn't bad enough, the IPCC wasn't set up as an *independent* scientific organization. Its mission statement reads: "Review by experts *and governments* is an essential part of the IPCC process."[3] [emphasis added]

Review by "experts" makes sense, depending on how one defines "expert." But IPCC findings being "reviewed by governments," i.e., politicians and government bureaucrats with political axes to grind, before being released to the public?

And "essential?" Why would "review by governments" be "essential" in reporting scientific research? Is "review by governments" "essential" in any other scientific discipline? And isn't another term for "review by governments" government censorship, or at least the very great risk of political interference in interpreting scientific results?

Of course, governments fund most scientific research, directly or indirectly. However, I can't imagine a government-funded physicist or biologist or chemical engineer, or indeed any other scientist with integrity, agreeing that the results of their research would be subject to "review by governments" before publication (unless, of course, they work directly for a government that, presumably, owns the research). This sort of government supervision of science belongs in a totalitarian system like the Soviet Union or Communist China.

Similarly, I can't imagine my former UVic English Department colleagues being happy if their academic work was "reviewed" by the government before publishing. I can't imagine my former journalistic colleagues being happy if their news and opinion articles were subject to "review by governments" before going online or in a publication.

In fact, I don't have to *imagine* how journalists would react to this kind of government interference. Here is Brodie Fenlon, editor-in-chief of CBC News, attacking Twitter's policy of adding "government-funded" to the network's Twitter feed (Twitter is now, of course, X), thereby (rightly I'd say) calling the CBC's political independence into question. Fenlon writes:

> Journalistic independence has been a central pillar of each of the five news organizations I've worked for ... and none more so than at my current employer, CBC News, where I am editor-in-chief.
>
> I haven't met a single journalist in all my years *for whom independence isn't a sacred virtue.* No matter the newsroom—print, magazine, digital or broadcast—journalistic independence is sacrosanct.
>
> *That's why reputable news organizations react so strongly to claims they are in the pockets of government, politicians or any other outside interest.* The idea is anathema to journalists, contrary to everything we believe and practise, and a threat to our public trust.4 [emphasis added]

And yet, although "review by governments" utterly destroys the integrity and independence of climate scientists, or at least the *perception* of their integrity and independence, alarmist climate scientists were willing to accept this scientific deal with the Devil.

Why? Because (I suppose) they believed that by exposing the evils of carbon dioxide they were saving the world, and therefore no price was too high. As Karl Popper has noted, "It is always flattering to belong to the inner circle of the initiated, and to possess the unusual power of predicting the course of history."5

Giving governments 'ownership' of climate policy

On the plus side, alarmist climate scientists also believed that getting governments involved in their process would make their findings more likely to be accepted and acted upon by governments.

"Will it be possible," asked Bert Bolin, the first IPCC chair, "to have climate policy without having a scientific group *in which various countries in the world have some political ownership*?"[6] [emphasis added]

Apparently climate research without "political ownership" is not possible, although this freedom from political ownership is the norm in other scientific disciplines. It doesn't hurt that getting cited and/or published by the IPCC is a career-booster, leading to more government grants (if government "reviews" your research, the least government can do is fund it as well).

How can any rational person fail to recognize that "review by governments" means there's pressure on scientists to toe a line and produce "research" that the governments doing the reviewing and funding want to hear (a.k.a., "consensus"/groupthink)?

(The IPCC's stance also raises the question: Why would democratic governments *want* to hear that fossil fuels, which underpin all modern economies, are sinful and need to be banned? Why would democratic governments *want* to make their constituents poorer, rather than better off? I'll try to explain this mystery in Chapter 12, to the extent that one can explain insanity.)

Of course, it may be true that carbon dioxide is an unmitigated evil and will lead to our doom. Unfortunately, the *process* used by the IPCC—"post-normal" science, emphasis on "consensus," suppression of differing viewpoints (where are the IPCC's minority reports? Surely at least *some* climate scientists disagree with the majority "consensus"!)—guarantees that the IPCC's findings will be tainted from the start, no matter now honest the scientists feeding information into the IPCC may be.

In accepting "review by governments" over its final conclusions, the IPCC drank from a poisoned chalice.

Biased conclusions in Summary for Policymakers

What does it mean that the scientists' conclusions are "reviewed" by governments? Blogger Donna Laframboise describes the process in her book *The Delinquent Teenager Who Was Mistaken for the*

World's Top Climate Expert. Her book does not present the IPCC process in a favorable light.

The full IPCC report itself is a compilation of research by more than 2,000 scientists and environmentalists and can run to thousands of pages; very few people will read the full report. What 99 per cent of politicians and the media, much less the public, read is the IPCC Summary for Policymakers for each report.

The Summary for Policymakers is the final stage of several drafts and revisions[7]; the final wording and what will be included in the Summary is hammered out line by line, over days of meetings, by government representatives (hence, "review by governments") and the "experts," including report authors but also, Laframboise observes, activist environmentalists who are not scientists themselves, many with "Greenpeace" or the equivalent in their resumes.[8]

This process is hardly the guarantee of "objectivity" and "transparency" that the IPCC promises in its mission statement; indeed, this process is the exact opposite of "objective."

Yet we, the public, are supposed to accept these reports as Gospel because there is a "consensus" (skeptics or dissenters need not apply).

But thanks to the way the Summary for Policymakers is produced, the final product is a highly politicized "consensus" that may (and sometimes does, as we'll see below) actually *differ* from the original report in key respects.

Reports revised to reflect Summary for Policymakers

The first (1991) IPCC report found *no evidence* that humans were warming the planet, much less dangerously, writing:

> The size of this warming [over the previous 100 years] is broadly consistent with predictions of climate models, *but it is also of the same magnitude as natural climate variability*. Thus the observed increase could be largely due to this natural variability; alternatively this variability and other human factors could have offset a still larger human-induced greenhouse warming. *The unequivocal detection of*

the enhanced greenhouse effect from observations is not likely for a decade or more.[9] [emphasis added]

So far so good. However, the alarmists didn't have to wait "a decade or more" to get the result they wanted.

The original version of the second IPCC report, released in 1995, also stated that "no study to date has positively attributed all or part [of the climate change observed to date] to anthropogenic causes."[10] So far, honesty preserved.

However, this and other sentences containing the same idea (no human influence detected) were *taken out* of the report and the following was inserted: "The balance of the evidence suggests a discernible human influence on global climate." This change was made, IPCC officials explained, "to ensure that it [the body of the report] conformed to a 'policymakers summary' of the full report."[11]

In other words, the baseline 1995 report was *revised to conform with the Summary for Policymakers,* which is the part of an IPCC report that is overtly politicized (i.e., "reviewed by governments").

In most scientific and business reports the summary reflects what's actually in the original document; the document isn't changed to conform to the summary, as the IPCC did. But, then, as we noted in Chapter 9 on "post-normal" science, the IPCC doesn't operate by the usual scientific standards.

Former IPCC president Rajendra Pachauri defended this cart-before-horse process as follows: "We necessarily have to ensure that the underlying report conforms to the refinements [i.e., "review" by experts and government]."

As climate blogger Donna Laframboise, who has extensively studied the IPCC's reports, puts it: "The IPCC tweaks their words so that *the underlying scientific sections accord with the version of reality that was hammered out by the politicians.*"[12] [emphasis added]

This, then, is "The Science" that we are expected to accept without questioning or criticism, "The Science" that tells us if we don't mend our carbon-using ways we will face "oblivion"; the reports that, hedged in "post-normal" language, offer instead of hard empirical facts a series of *observations* and *speculations* about aspects of climate, modulated by terms like "medium confidence," "very likely," "unlikely," and so on, as noted in Chapter 9.

Reports themselves don't prophesy doom

There is some irony here, by the way. The IPCC reports in themselves don't prophesy doom, at least not directly.[13] Indeed, the 2001 report specifically stated:

> In climate research and modeling, we should recognize that we are dealing with a coupled non-linear chaotic system, and therefore that *long-term prediction of future climate states is not possible*. The most we can expect to achieve is the prediction of the probability distribution of the system's future possible states by the generation of ensembles of model solutions.[14] [emphasis added]

Then these non-"long-term predictions" that are "not possible" are morphed by the Summaries for Policymakers into confident long-term predictions of doomsday, based on which we are expected to completely remake our civilization, as we'll see in Chapters 15 and 16.

Based on the summaries, UN representatives take up the megaphone to warn us of "oblivion" and "thin ice" that is not actually predicted by the report, doom-saying that is further magnified by the mass media and politicians. This is the hyping process that, as Steven Koonin noted, occasionally "embarrasses" climate scientists.[15]

It's sad, I think, that these "embarrassed" scientists don't have the courage to stand up as a group or individually and state publicly that the Summaries for Policymakers and the inflammatory remarks by UN Secretaries-General Ban Ki-Moon ("oblivion") and Antonio Guterres ("code red for humanity"), among others, are exaggerated and offer a correction.

But only a tiny minority speaks up lest, one can only guess, they appear to break the "consensus," become known as "deniers," and are "denied" funding and academic positions.[16]

Later IPCC reports (slightly) less alarmist

Meanwhile the IPCC reports come out, every five years or so, and, oddly, they have become *less* inflammatory over the last decade or so.

For example, the "catastrophic" predictions magnified by the UN, politicians and the media are based on two of five "scenarios" used by the IPCC to model different approaches to the "climate crisis." These two extreme scenarios assume some form of "business as usual," with emissions growing, leading to warming (it is claimed) of up to 6°C.

However, the two most recent reports (AR5 and AR6) have acknowledged these two extreme scenarios are unlikely and predicted lower possible future temperature increases—temperatures well within humanity's ability to adapt to, by the way. The most recent report, AR6 (2021), flatly states: "High-end emission scenarios ... are becoming less likely with climate policy and technology change."[17]

Then "governments" and "experts" (including activist environmentalists) produced an AR6 Summary for Policymakers based on which, for Guterres and the media, the less-scary non-predictions become "code red", feeding to the public "facts" that are untrue and/or exaggerated to make us meekly accept carbon taxes, a lower standard of living, and less personal freedom.

The IPCC process is so flawed it is astonishing anyone takes the organization seriously, but, alas, the IPCC *is* taken seriously and its "conclusions," hyped by the UN, politicians and the media, promise nothing but misery for humanity.

We'll discuss the disastrous economic effects of the "Net Zero by 2050" policy, a product of IPCC doom-mongering, in Chapter 14. But first, in Chapter 12, we'll examine why democratic politicians, of all people, support the insane Net Zero policy, which can only make their citizens poorer, despite the lack of any rational evidence for its existence.

Notes to Chapter 11

---

[1] IPCC, "History of the IPCC." Available at https://www.ipcc.ch/about/history.

[2] IPCC, "16 Years of Scientific Assessment in Support of the Climate Convention," December, 2004, p. ii. Available at https://www.ipcc.ch/site/assets/uploads/2019/03/16th-anniversary-brochure.pdf

[3] IPCC, "16 Years...", p. ii.

[4] Brodie Fenlon, "Twitter changes had to be called out." *National Post*, April 21, 2023.

[5] Karl Popper, *The Open Society and Its Enemies, Vol. 1*. Princeton, NJ: Princeton Univ. Press, 1971 (1943), p. 4.

[6] Quoted in Stephen Schneider, *Science as a Contact Sport*. Washington, DC, National Geographic, 2009, p. 125. Schneider's book offers an account of the infighting around the founding of the IPCC and what occurs in Summary for Policymaker meetings, including politicians and/or government bureaucrats very much inserting themselves in the process.

[7] For the IPCC's description of the writing and review process, see https://www.ipcc.ch/review-and-scoping-process-ar6-wg1.

[8] Donna Laframboise, Chapter 6: "Activists," *The Delinquent Teenager Who Was Mistaken for the World's Top Climate Expert*. Ivy Avenue Press, 2011, Kindle edition.

[9] IPCC, 1991, "Policymaker Summary of Working Group I (Scientific Assessment of Climate Change), p. 64. Available online.

[10] S. Fred Singer, Letter to IPCC (Working Group I) Scientists, 1998, available at http://web.archive.org/web/19980629122454/http:/www.sepp.org/ipcccont/ipccflap.htm

[11] IPCC, Climate Change 07, Working Group I: The Physical Basis of Climate Change, Chapter 9, p. 704.

[12] Laframboise, *Teenager,* Chapter 22, "Science Has (Not) Spoken.".

[13] From Michael Shellenberger, *Apocalypse Never: Why Environmental Alarmism Hurts Us All*. Harper, 2020, p. 4: "The IPCC did not say the world would end, nor that civilization would collapse, if temperatures rose above 1.5 degrees Celsius."

[14] IPPC, Third Assessment Report (2001), Section 14.2.2.2, page 774.

[15] Steven Koonin, *Unsettled: What Climate Science Tells Us, What It Doesn't, and Why It Matters*. Dallas: Ben Bella Books, 2021, p 4.

[16] One climate scientist did speak up, hurricane expert Christopher Landsea. In 2005 he quit the IPCC over false claims that "global warming" was making hurricanes more severe. Landsea is, alas, a rarity.

[17] IPCC, AR6, 2021, Working Group III, Section 3.2.1.1 "Reference scenarios." Available online.

Chapter 12: Science? Or political ideology?

"No, no!" said the Queen. "Sentence first—verdict afterwards."

It's puzzling why democratically elected politicians, who are supposed to have the good of their constituents at heart, have allowed themselves to be seduced by alarmist climate science's vision of catastrophic global warming.

It's an even greater mystery why they have been seduced by the "Net Zero by 2050" vision for "solving" the supposed global-warming problem. This "solution" will devastate (and is already devastating) the global economy, which means the lives and livelihoods of literally billions of people.

Of course, the politicians must be telling themselves that they are *saving* civilization from catastrophic warming (as former British prime minister Boris Johnson was persuaded, see Chapter 8), which even the less-alarmist climate scientists don't believe (as we've seen, some are even "embarrassed" by the wilder claims, although they're not doing anything about it).

If this is so—if politicians believe they are planetary saviors—then they haven't *actually examined* the evidence outside the highly politicized offerings of the IPCC's Summary for Policymakers, as magnified by the UN and the media.

Independently examining the evidence before creating government policy is called "due diligence" and should include a "cost-benefit" analysis of the proposed Net Zero measures. This basic vetting, an indispensable requirement for government decisions, has clearly not been done by alarmist-influenced political leaders.

In Chapter 14 we'll look at what basic due diligence and a cost-benefit analysis would tell our political leaders about the (potentially disastrous) economic effects of Net Zero by 2050. For now, we'll try to understand why politicians would support a policy that will bring economic misery to their nations' voters, and why, *for the first time in history,* liberal democratic politicians are *deliberately trying to make their citizens poorer, rather than better off.*

Al Gore started the climate bandwagon

We get some clues if we look at the early supporters of anti-carbon measures, from the late 1980s to the 2000s, typified by Al Gore, a U.S. senator, then vice-president in the Bill Clinton administration, and then a losing presidential candidate.

We've already seen in Chapter 7 that, as early as 1992 in his book *Earth in the Balance*, Gore wanted the world's nations to jump on the global-warming bandwagon basically on his say-so, without any actual empirical evidence amounting to scientific proof[1]—even the first IPCC report in 1991 hadn't found a human "signal," as we saw in Chapter 11.

Nothing daunted, Gore argued that scientific proof was overrated if there was any possibility that his doomsday predictions could be true (this is known as the "precautionary principle"—if something *might* be disastrous, we must act as if it *will be* disastrous).

Gore's anti-scientific approach got a surprising amount of support. For example, economist Andrew Brod wrote: "Fortunately, people finally seem to understand the fallacy of requiring proof"[2]—i.e., proof that humans are the *prime cause* of global warming and that carbon dioxide is the "control knob" of climate.

The alarmist climate claims offer sweeping predictions with massive, and possibly very negative, ramifications for our society, but who needs actual, like, you know, proof? Especially since we're dealing with "post-normal" science here where standards of "proof" are elastic.

Taking a more rigorous position on what constitutes good "science," astronomer Carl Sagan observed, "Apocalyptic predictions require, to be taken seriously, higher standards of evidence than do assertions on other matters where the stakes are not as great."[3] Note: *higher standards of evidence* for "apocalyptic predictions," a.k.a., a *higher* standard of "proof."

So what are the "higher standards of evidence" that support these apocalyptic climate predictions with so much at stake (nothing less than the future of Western civilization)? Where is the empirical confirmation, the much-despised "proof"?

It isn't there. Instead, alarmist political supporters grasp at ideology and computer models to support their predictions. Here are three oft-cited examples.

Christine Stewart, environment minister in the Jean Chrétien Liberal government, in 1998 declared she didn't care a whit about proof: "*No matter if the science is all phony,* there are collateral environmental benefits."[4] [emphasis added]. This repudiation of scientific honesty came from a senior Canadian government minister.

A senior American government official and a major force behind the Kyoto Accord, Timothy Wirth, had this to say in 1990: "We have got to ride the global-warming issue. *Even if the theory of global warming is wrong,* we will be doing the right thing in terms of economic policy and environmental policy."[5] [emphasis added]

And Richard Benedick, a U.S. deputy assistant Secretary of State, put the anti-scientific case very clearly in 1989: "A global-warming treaty must be implemented *even if there is no scientific evidence to back the [enhanced] greenhouse effect.*"[6] [italics added]

I can't imagine these statements being uttered about any other scientific discipline, such as physics: "Even if the theory of relativity is wrong, we should believe in it anyway." Or medicine: "Even if cigarettes don't cause cancer, claiming they do is the right thing to do if this belief will get people off tobacco."

Alarmists' ultimate goal: Destroy the capitalist system

These three climate statements only make sense in the climate context *if it is the IPCC's job is to provide scientific cover for what some politicians and bureaucrats wanted to do anyway*, which appears to be to sacrifice the economy to save the environment, but with the ultimate aim of creating a socialist/communist social system.

Maurice Strong, a Canadian who was one of the leading founders of environmentalism (he organized the first Earth Summit in Rio de Janeiro in 1992), makes this connection of politics and environmentalism crystal clear: "We may get to the point where *the only way of saving the world will be for industrialized civilization to collapse*."[7] [emphasis added]

The Club of Rome, an alarmist environmental think-tank, in 1972 published *Limits to Growth,* which predicted widespread shortages of key industrial minerals and food in the coming decades (so far, none of their predicted global shortages have occurred).

In 1991 the Club published *The First Global Revolution,* which includes the following:

> The common enemy of humanity is man. *In searching for a new enemy to unite us,* we came up with the idea that pollution, the threat of global warming, water shortages, famine and the like would fit the bill. All these dangers are caused by human intervention, and it is only through changed attitudes and behavior that they can be overcome. *The real enemy then, is humanity itself.*"[8] [emphasis added]

In other words, the Cold War ends in 1991 and the elites decide we, the public, need a new "enemy"—shades of EngSoc in Orwell's *1984*!—to keep us nodding our bobble-heads with whatever the "official line" happens to be at any given time.

But that's okay because we ("humanity") are being bamboozled ("searching for a new [imaginary] enemy to unite us") *for our own good*. (The Club of Rome's approach, by the way, is a fine example of a solution seeking a problem.)

But why use deception? Why not just be straight with the public and say: "We're going to lower your standard of living, for your own good. What do you think?"

Somehow I doubt the average Canadian or American or European (or citizen anywhere) would vote to reduce his/her standard of living, as we're being asked to do, without this verbal smokescreen.

In the same vein, biologist Paul Ehrlich, author in 1968 of *The Population Bomb* that predicted massive global famines by the mid-1980s, (didn't happen; in fact, fewer people are starving today, proportionally, than ever before in human history), wrote in 1973:

> A massive campaign must be launched *to de-develop the United States*. De-development means bringing our economic system into line with the realities of ecology and the world resource situation.[9] [emphasis added]

Ehrlich, a neo-Malthusian thinker who completely misread the population/starvation equation in 1968, nonetheless feels confident in telling us that we should accept a much lower standard of living today based on, basically, nothing but his say-so.

And, of course, he and his elite cronies will also decide the point at which we, the great unwashed, are finally "in line with the realities of ecology" and take the pressure off. Which, based on past experience with totalitarian societies, is never.

Or, as Orwell put it in *1984:* "If you want a picture of the future, imagine a boot stamping on a human face—for ever."[10]

That said, the above comments give us some insight into why politicians in the democratic countries are so willing to abandon rationality and seek to destroy the current economic system, based on manufactured fears of some vaguely defined future disaster that may never happen and may not even exist as a problem outside of computer models.

Environmentalism as a secular religion

Environmentalism has been a secular religion for at least five decades now. We are all (or almost all) "environmentalists," in that we all care about the quality of our natural surroundings in a way that the earlier "Robber Baron" capitalists did not (at least in the alarmist imagination).

And that means politicians must also be (or appear to be) environmentalists if they want to get elected and keep getting re-elected. And since "anthropogenic climate change" is being presented to the public and politicians as an *environmental* issue, and has captured much public support (hands up everyone who is *in favor of* a climate apocalypse), even skeptical politicians are forced to jump on the anti-global-warming bandwagon as an "environmental" issue. It doesn't matter that the bandwagon's proposed "solutions" make no scientific or economic sense.

It's ironic that Margaret Thatcher, former Conservative (and conservative) prime minister of Great Britain, was a strong supporter of the creation of the IPCC in 1988, ironic because she was also a strong supporter of Western civilization.

Her purpose wasn't to demolish Britain's economic system, like modern IPCC supporters; it was to modernize the economy, in part by reducing the nation's dependence on coal, an industry that was government-owned, a major money-loser, and a major polluter.

Naturally, Britain's 200,000 mineworkers weren't happy about this policy and in March 1984 the Marxist-led National Union of Mineworkers called a strike that lasted almost a year before giving up. Thatcher was then able to privatize and eventually close most of the coal pits.[11]

But why her attack on the coal industry? To reduce carbon emissions from coal? In 1988 carbon emissions as a source of "global warming" were a consideration for Thatcher, but her ultimate goal was to promote nuclear energy. "Nuclear power production did not produce carbon dioxide," she wrote, "nor did it produce the gases that led to acid rain. It was a far cleaner source of energy than coal."[12]

But she faced a major problem. In 1979 there was a partial core meltdown at the Three Mile Island nuclear plant in Pennsylvania. The damage was minimal but, coincidentally, that same year saw the release of *The China Syndrome,* a popular film about the chance of a much more serious core meltdown (melting through the Earth to China, as it were).

As a result, Britons and many other nations were scared off nuclear energy. Thatcher hoped that the IPCC's prestige would make nuclear more acceptable to her citizens as a substitute for coal.

No such luck. She quickly discovered that her IPCC initiative "did not attract the environmental lobby towards it [nuclear power]; instead they used the concern about global warming to attack capitalism, growth and industry."[13] In her 2002 book *Statecraft* she noted:

> The doomsters' favorite subject today is climate change. This has a number of attractions for them. First, *the science is extremely obscure so they cannot easily be proved wrong*. Second, we all have ideas about the weather: traditionally, the English on first acquaintance talk of little else. Third, since clearly no plan to alter climate could be considered on anything but a global scale, *it provides a marvelous excuse for worldwide, supra-national socialism.*[14] [emphasis added]

She was also astonished at the speed with which "the new dogma about climate change has swept through the left-of-centre governing classes."

Marxists hide behind environmentalism

But in reality, this adoption of "global warming" as the latest leftist fad isn't hard to explain. Just as the IPCC was building up steam for climate alarmism in the early 1990s, the Marxist world was in crisis. In 1989 the Berlin Wall fell and in 1991 Soviet Russia imploded. Marxism had apparently run its political course, clearly revealed as a complete failure, and the Marxist Left in the developed world was fighting for its very survival.

No problem! Marxism quickly latched onto the environmental movement in general, "global warming" in particular, as its new rationale for undermining the hated capitalist system, with the IPCC as its Trojan Horse.[15] (One way of describing these Marxist opportunists, by the way, is "watermelons"—green on the outside, red on the inside.)

It was this undermining of capitalism and the move toward global socialism under cover of the IPCC that Thatcher belatedly recognized.

Unfortunately, far too many citizens in Canada, the United States and Europe still do not understand that what appears to be an existential fight against "global warming"/"climate change" is just a convenient cover for the Marxist Left's unceasing efforts to save itself while destroying capitalism, even though, ironically, capitalism has been very successful at providing a good material life for most of the "working classes" in the developed world.

If "global warming" is, in fact, mainly the excuse for a Marxist undercover operation, we can be, and should be, very suspicious of the "reality" of this so-called "climate crisis."

As further evidence of the ephemeral nature of "global warming," the former head of the IPCC, Rajendra Pachauri, in defending his institution in 2010 after Climategate, noted: "If the IPCC wasn't there, *why would anyone be worried about climate change*?"[16] [emphasis added]

It's an interesting question, isn't it? If we didn't have the IPCC and other doomsayers, would we be worried about "global warming"? Would we have people believing they are going to go "extinct"? That the future is gloomy and even hopeless, as many young people (including, shamefully, children) now believe?

Or would we just go on living our lives as usual while coping, as we have for thousands of years, with the usual floods, storms, forest fires, droughts, etc. (a.k.a., "extreme weather"), which have always existed and always will?

So, is catastrophic "global warming" a real thing? Or mostly a tool for an anti-capitalist ideology?

Climate alarmism's Vision of the Anointed

In 1995, before "global-warming" alarmism was in full flight, American thinker Thomas Sowell offered a useful insight into the Club of Rome's ideological process of "searching for an enemy to unite us" using environmentalism/climate change as the enemy.

In his 1995 book *The Vision of the Anointed,* Sowell identified environmental zealots (or any zealots) as believing themselves "The Anointed." The Anointed, he wrote, believe they have a special, God-given (even if they don't believe in God) mission to save the world, or make it permanently better in some way.

The Vision of the Anointed offers "a special state of grace for those who believe in it." Those who follow this vision "are deemed to be *not merely factually correct but morally on a higher plane*" [emphasis added], while those who don't accept this vision—the "benighted"—"are seen as being not merely in error, but in sin."[17] The most recent incarnation of the Anointed mind-set is, of course, "the woke."

Sowell identifies the political program of the Anointed as follows (and again, this was written in 1995, *before* "global warming" became A Thing):

1. Assertions of a great danger to the whole society, a danger to which the masses of people are oblivious.
2. An urgent need for action to avert impending catastrophe.
3. A need for government to drastically curtail the dangerous behavior of the many, in response to the prescient conclusions of the few.
4. A disdainful dismissal of arguments to the contrary as either uninformed, irresponsible, or motivated by unworthy purposes.[18]

Sound familiar? Doesn't this Anointed political program fit, to the letter, the public stance of the climate alarmists of the world? And how many of these allegedly "great dangers" have actually proven to be real? Almost none.[19]

There's no reason why the current "climate emergency" is any different—it is ultimately a political program of the Anointed (another Marxist guise) and is not, as claimed, based in "science."

Affluence brings more care for the environment

One of the ironies of environmentalism's anti-capitalism is that the more affluent a society is, the more its citizens seek to protect the environment. In their 2007 book *Break Through,* Michael Shellenberger and Ted Nordhaus call this protective mind-set "post-materialism." They write:

> Environmentalism and other progressive social movements of the 1960s were born of the prosperity of the postwar era and the widespread emergence of higher-order post-materialist needs. *As Americans became increasingly wealthy, secure, and optimistic, they started to care more about problems such as air and water pollution and the protection of the wilderness and open space.*
>
> This powerful correlation between increasing affluence and the emergence of quality-of-life and fulfillment values has been documented in developed and undeveloped countries around the world.[20] [emphasis added]

They add, and in contrast to what the environmental movement wants us to believe: "The satisfaction of the material needs of food and water and shelter is not an *obstacle to* but rather the *precondition for* the modern appreciation of the nonhuman world."[21] Poor people can't afford to care for the environment; affluent people are willing to pay for a better environment.

But the current Net Zero ideology will, in fact, make us poorer (*much* poorer), not better off, as we'll see in Chapter 14. That means, inevitably, that while the environment has actually been *improving* under our current capitalist affluence, the impoverishing policies of the alarmists will damage the environment.

This is the direction in which we are blindly being led by the current crop of Net Zero politicians under the guise of "environmentalism" and "sustainability." They will succeed unless we, the voters, tell the alarmist politicians in no uncertain terms that increased poverty and environmental degradation is not the way we want to go.

We'll discuss in Chapter 15 how fear of "global warming" and the Net Zero approach to solving this (possibly non-existent) "problem" will, in fact, make us poorer—a lot poorer.

But for now, let's stay alert to the idea that "global warming" may not be a *real* threat, but mostly a rhetorical political club wielded by the revitalized Marxist Left (the Anointed on steroids) against its avowed enemy: capitalism and affluence.

And we should be alert to the idea that under the umbrella of "environmentalism," many of our politicians are, possibly unwittingly, possibly in full consciousness, playing their part as Lenin's "useful idiots."

Notes for Chapter 12

---

1 Al Gore. *Earth in the Balance: Ecology and the Human Spirit.* Toronto: Penguin Books, 1993 (1992), p. 39. Gore's words, which we also saw in Chapter 7 : "If, when the remaining unknowns about the environmental challenge enter the public debate, they are presented as signs that the crisis may not be real after all, it undermines the effort to build a solid base of public support for the difficult actions we must soon take."

2 Andrew Brod, "It's Not Too Late—Yet—to Counteract Global Warming." *Greensboro News & Record,* April 29, 2007.

3 Carl Sagan, "Nuclear War and Climatic Catastrophe: Some Policy Implications," *Foreign Affairs,* Winter 1983/84, pp. 257-258.

4 Quoted in Peter Menzies, "Concerned about the cost of Kyoto." *Calgary Herald,* Dec. 14, 1998, p. A12.

5 Quoted in Robert James Bidinotto, "Environmentalism: Freedom's Foe for the 90s." *The Freeman,* vol. 40, no. 11, November 1990.

6 Richard Benedick, "Who needs evidence?" 1989. Quoted in Dixy Lee Ray, *Trashing the Planet.* New York: Harper Perennial, 1990, p. 167.

7 Maurice Strong, quoted in *National Review,* Sept. 1, 1997.

8 Alexander King & Bertrand Schneider, *The First Global Revolution: A Report by the Council of the Club of Rome.* New York: Simon & Schuster, 1991, p. 85.

[9] Paul & Anne Ehrlich, John Holdren, *Human Ecology: Problems and Solutions*. W.H. Freeman, 1973, p. 4.

[10] George Orwell, *1984*. Penguin Books, 2008 (1949), p. 280.

[11] Sarah Pruitt, "When Margaret Thatcher Crushed a Miner's Strike." *History Channel*, Nov. 16, 2020. Available online.

[12] Margaret Thatcher, *The Downing Street Years*. London: HarperCollins, 1979, p 640.

[13] *Ibid.*, p. 640.

[14] Thatcher, *Statecraft: Strategies for a Changing World.* New York: Harper, 2002, pp. 449-450.

[15] Canadian investigative journalist Elaine Dewar describes the process of left-wing infiltration of the environmental/climate-change movement as a "cloak of green" in her book *Cloak of Green: The Links Between Environmental Groups, Government and Big Business*, Lorimer, 1995.

[16] "Climate Science Leader Rajendra K. Pachauri Confronts the Critics." Interview with *Science*, Jan. 29, 2010.

[17] Thomas Sowell, *The Vision of the Anointed: Self-Congratulation as a Basis for Social Policy.* New York: Basic Books, 1995, pp. 2-3.

[18] *Vision of the Anointed,* p. 5.

[19] See "Earth Day at 52: None of the eco-doomsday predictions have come true." *Net Zero Watch,* April 22, 2022. Available online.

[20] Ted Nordhaus & Michael Shellenberger, *Break Through: From the Death of Environmentalism to the Politics of Possibility*. Boston: Houghton Mifflin, 2007, p. 6.

[21] Nordhaus & Shellenberger, *Ibid.*, p. 27.

Chapter 13: Doomsday? Or a more prosperous future?

When I used to read fairy-tales, I fancied that kind of thing never happened, and now here I am in the middle of one!

The United Nations is a huge, global organization, so it's not surprising that sometimes the left hand doesn't know what the right hand is doing.

When the United Nations is wearing its IPCC "global-warming-apocalypse" hat, we are warned of climate "oblivion" ahead if we keep emitting the current levels of carbon dioxide as an fossil-fuel-based civilization.

When the United Nations is wearing its "human development/Gross Domestic Product (GDP)" hat, however, it actually promises a much brighter future for humanity, at least economically.

And, curiously, this bright economic future is *especially* likely if we continue with a "business-as-usual" approach that doesn't try too hard to curb our carbon emissions. How can we square these two diametrically opposite, indeed, incoherent, positions?

The IPCC predictions of global-warming disaster in the Sixth (2021) report (most enthusiastically in the Summary for Policymakers) are based on five so-called Shared Socioeconomic Pathways (SSP) scenarios (called Representative Concentration Pathways or RCPs in the fifth report). They are:

- SSP5: "fossil-fuel development" (carry on using fossil fuels for economic growth, also known as "business as usual") (= RCP8.5);
- SSP4: "inequality" (= RCPs3.4-6);

- SSP3: "regional rivalry" (= RCP7);
- SSP2: "middle of the road" (= RCP4.5); and
- SSP1: "sustainability" (= RCP2.6 and 1.9).[1]

Figure 13.1 shows how much these scenarios are expected to increase global temperatures (using the RCP terminology).[2]

RCP	Forcing	Temperature	Emission Trend
1.9	1.9 W/m2	~1.5 °C	Very Strongly Declining Emissions
2.6	2.6 W/m2	~2.0 °C	Strongly Declining Emissions
4.5	4.5 W/m2	~2.4 °C	Slowly Declining Emissions
6.0	6.0 W/m2	~2.8 °C	Stabilising Emissions
8.5	8.5 W/m2	~4.3 °C	Rising Emissions

Figure 13.1: IPCC RCP scenarios and expected warming. Source: Sense: "Making Sense of Climate Change Scenarios for Investors: Primer to Climate Scenarios"

The RCP numbers (8.5, 2.6, etc.), by the way, represent "radiative forcing," which is another way of describing the amount of increased global heating each scenario is supposed to create, expressed in Watts per square metre (W/m^2).

You can read more about radiative forcing and how the RCP numbers are misleading (not surprisingly, given that this is the IPCC) in **Sidebar 6: Radiative Forcing** at the end of this chapter.

The RCP8.5 doomsday scenario

As noted above, the doomsday scenario is RCP8.5, and indeed, through the various IPCC reports, the RCP8.5 scenario with its predictions of doom is the one that produces the scary headlines, and it is also a scenario that alarmists have in the past considered highly likely.[3] For example, a climate scientist at a recent social event I attended predicted that none of the children at the gathering would reach the age of 90. Wow! That's doomsday-scenario alarmism!

However, in its AR6 (2021) report, the IPCC now accepts that "high-end emission scenarios, such as RCP8.5 and SSP5-8.5 (in the SSPs, the number after the hyphen describes the amount of radiative forcing in W/m^2) are becoming *less likely* with climate policy and technology change."[4] [emphasis added]

One reason for this downgraded end of the world, for the IPCC, is that "technological change" now includes less reliance on coal. (Apparently the IPCC is unaware of China's and India's increasing reliance on coal-fired power plants—more on that in Chapter 16).

In other words, the latest IPCC report no longer promotes the extreme and scary RCP8.5 and SSP5-8.5 scenarios as plausible.[5] This means that, in the IPCC's crystal ball, we're going to get warmer, but not sauna warmer—a mere 2-3°C perhaps. Whew! That's a relief! Not that these extreme 8.5 predictions were ever plausible, but it's nice to know the IPCC now acknowledges this.

There is a fly in the ointment, however. You may recall from Chapter 11 that even though the claims in the IPCC reports themselves are (relatively) restrained, the Summary for Policymakers reports are written by "governments" and (mainly) activist environmentalists with the express aim of creating climate fear so we'll all fall into line.

And so, right on cue, just as the AR6 report itself downplays the extreme SSP5/RCP8.5 scenarios, the Summary for Policymakers for the IPCC's 2023 "Synthesis" report brings them back with the same old alarmist vengeance, to wit:

> Climate change is a threat to human well-being and planetary health (very high confidence). *There is a rapidly closing window of opportunity to secure a livable and sustainable future for all* (very high confidence).[6] [emphasis added; note that the "very high confidence" of doom isn't in the original report]

This is yet another case where the Summary for Policymakers goes well beyond the actual report—and also well beyond "The Science".

Why we will be better off

Now—and here's where the UN's right hand doesn't know what the left is doing—it's a curious and paradoxical fact that numerous studies by the UN and other organizations show that humanity would be materially a *lot better off* by 2100 under the "extreme," "business as usual" SSP5/ RCP8.5 scenarios than under the "less-extreme" scenarios in which we more or less bid farewell to fossil-fueled civilization.

The world may be warmer by 2-3°C or more under SSP5/RCP8.5, perhaps, but we'll also be *much richer and thus much more capable of handling the increased warmth* (if, indeed, this greatly increased warmth even occurs).

But how could we possibly be better off if the world is warming catastrophically?

The reason is that despite "global warming," if we look at past global economic performance and we don't have another world war (or economic collapse caused by the current disastrous Net Zero policies), the global Gross Domestic Product is still expected to increase by around US$6-trillion a year, give or take, at least as far as 2060.[7]

Global GDP is going up, not down

The Organization for Economic Co-operation and Development (OECD) reports that in 2020 world GDP was US$100-trillion (a trillion is a thousand billion), and is expected to rise by US$234-trillion to more than $300-trillion by 2060.[8]

In other words, by 2060 the world's people—developed and developing—should be *three times richer* than they are now *if we carry on with business more or less as usual.* Extrapolating another 40 years to 2100, we'd expect the world's people to be at least 5-6 times richer. Here's where the IPCC's SSP scenario data becomes interesting.

First, to backtrack a bit: why does each generation become better off than the previous one, barring war or natural calamities like earthquakes? Because each generation benefits from the advances of the previous generation.

We, the current generation, don't have to build the cities, the electricity grids, the highways, the railroads, the medical systems, and so on; our parents and grandparents bequeathed these treasures to us.

We don't have to re-invent the wheel, as it were, or the personal computer; each previous generation's inventions come more or less for free as a baseline for the current generation to *add to*, and so later generations are almost always better off.

With this in mind, let's look at GDP growth in percentage terms. The McKinsey Global Institute (which takes an alarmist view of global warming) expects the global GDP to increase about 2 per cent a year for the next 50 years despite, presumably, the predicted "catastrophic" climate change.

Let's pretend the global GDP per person (i.e., income) is US$100 in 2020. Assuming a GDP increase of two per cent a year at compound interest, and that inflationary pressure is accounted for, what would this individual's income be in 80 years, by the year 2100?

Our trusty Excel spreadsheet shows us that by 2100, instead of having $100 a year to spend, the average person would have $487 a year. Our spending power has increased almost five times! We can buy almost five times as much "stuff" (again, all things being equal and inflation accounted for)!

And this increase assumes fairly low growth compared to, say, historical growth of almost 4 per cent from 1950-2014.[9] This potential growth should be good news for most of us, although of course not to the gloomy neo-Malthusians who think we're so rich already that we're destroying the planet.[10]

'Business as usual' means higher incomes

Let's suppose we abandon the climate alarmism of Net Zero by 2050 and accept a more "business as usual" approach to economic growth, such as the IPCC's SSP5 scenario.

This means we continue to use fossil fuels to make the world economically better off while *gradually* moving off fossil fuels that will, eventually, become scarce and therefore prohibitively expensive to extract. In that case, the increase in GDP is, say, 4 per cent a year,[11] which is more like the norm for the past seven decades.

At 4 per cent a year, with compound interest, a GDP of $100 in 2020 will become $2,304 by 2100! Our spending money has increased more than twentyfold!

If we assume the starting point for this 4 per cent annual growth is the current global GDP of $100-trillion, then with a population of eight billion, each person's share is about $12,000. If population stays at eight billion, and with 4 per cent growth per year, over 80 years the average person's share increases to $266,000!

So even though, according to the AR6 Summary for Policymakers, this level of growth would be "a threat to human well-being and planetary health"—a drought-ridden, flooded, blasted planet, basically—somehow we're much, much better off. Weird.

Figure 13.2 shows the relative increases of GDP per person under the five SSP scenarios based on OECD projections[12]:

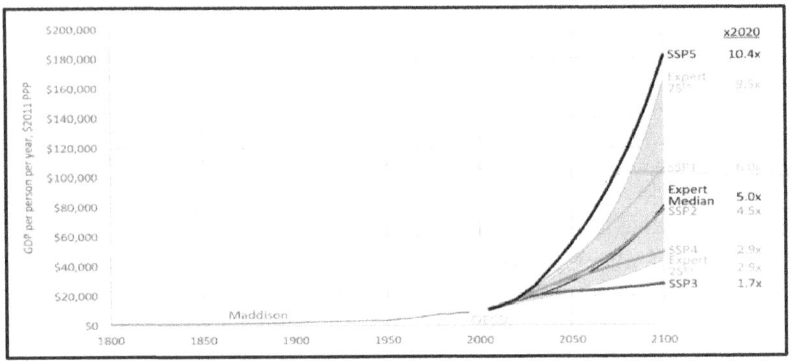

Figure 13.1: Global GDP growth per capita from 1800-2100 under the five IPPC SSP scenarios. Source: Bjorn Lomborg, "Welfare in the 21st Century"

Note, again, that GDP growth per capita in the SSP5 (RCP8.5) scenario is *10.4 times* GDP in 2020. And this is the scenario that is going to lead us to Thermageddon!

SSP1 is the IPCC's Utopia and, like most utopias, it sounds too good to be true and therefore, like most utopias, is highly prone to going wrong (Plato's *Republic*, for example, the prototype for most Western utopias, would have been a totalitarian dictatorship).

The other three SSP scenarios are SSP2 ("middle of the road"), SSP3 ("regional rivalry"), and SSP4 ("inequality"). They produce less long-term growth in GDP than SSP1 or SSP5.[13]

Now, if we assume that catastrophic climate change is a Real Thing (it isn't, but assuming), we will obviously be facing more severe hurricanes and flooding and forest fires and rapid sea-level rise and "extreme weather"—the horrific picture of our planet that presenter David Attenborough paints in his *Climate Change: The Facts* documentary (noted in Chapter 10).

It's obviously going to cost a lot of money and resources to deal with these problems. But how much? Nobody really knows, but it's very possible that, if we abandon Net Zero and don't drastically cut off fossil fuels, we'll have more than enough wealth to deal with whatever climate/weather problems crop up (as we have being doing anyway, long before doomsday "global warming" became popular).

Adaptation vs. mitigation

For climate alarmists, the cost of *not* taking extreme climate action will be enormous and could cause social and environmental collapse as we overuse and then run out of key resources.[14]

Alarmists believe it will cost us *less*—even *much less*—to mitigate (stop) climate change than trying to *adapt* to climate change. As climate change worsens, they believe, we will find ourselves buried under an avalanche of calamities (fires, floods, hurricanes, droughts, etc.) while running out of resources, and our increased GDP won't help us. Unless we *stop* climate change, we're doomed!

For climate realists, like Danish political scientist Bjorn Lomborg (the Skeptical Environmentalist), the picture is considerably less gloomy—*if* we don't undercut our current economic prosperity with ill-considered, rapid Net Zero anti-carbon policies.

Indeed, we are already seeing the negative results of these Net Zero policies in Europe, which is generally ahead of North America in adopting the "green" ideology.

Life in European nations that have embraced Net Zero has become considerably (sometimes absurdly) more expensive for the average household, especially for electricity (home heating and the like). Just one example: electricity prices for European households are 80 per cent above other nations in the G20 economic group.[15]

What climate doomsters miss, Lomborg notes, is that we're aren't going to sit twiddling our thumbs while the climate problems accumulate and eventually overwhelm us (for some ideas about how we could cope, see Chapter 17).

What's absent in alarmist predictions, Lomborg says, "is the fact that adaptation to the problems is neglected, *leading to climate costs being vastly exaggerated* compared to the more realistic costs of a world where adaptation takes place."[16] [emphasis added] In other words, a stitch in time will save nine.

SSP scenarios all offer growth

Lomborg notes that all five SSP scenarios offer at least some GDP growth, with SSP5 ("fossil-fueled development") the most growth (10x), and SSP1 ("sustainability") showing acceptable growth (6x) in average GDP per person.

At the same time, Lomborg notes, in none of the SSP scenarios do we even approach "Net Zero" in terms of full reliance on "sustainable/renewable" energies, as Figure 13.3 shows:

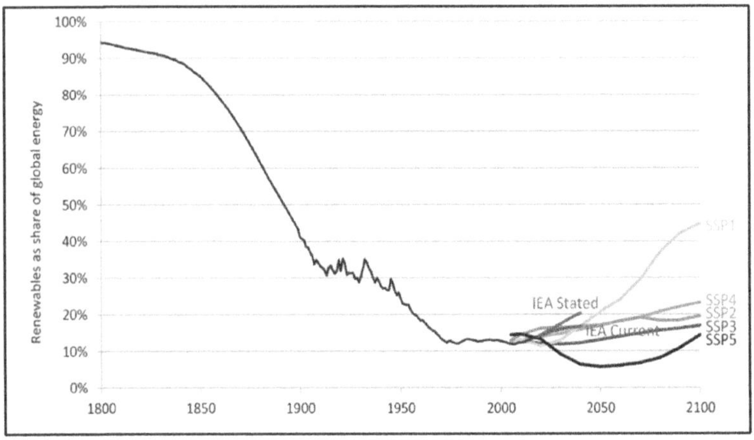

Figure 13.3: Percentage of 'renewable' energy use under the five SSP scenarios. Source: International Energy Agency, in Lomborg "Welfare in the 20th Century"

Even the "sustainable" SSP1 scenario can count on only 45 per cent "renewable" energy by 2100; the remaining energy will have to come from fossil fuels and/or nuclear and/or some energy source so far undreamed of.[17]

The take-home message is that *renewable energy sources, like wind and solar, cannot sustain a complex technological civilization,* and Net Zero by 2050 (or earlier) is a fantasy or, more accurately, a dangerous delusion.

It's worth noting that Lomborg accepts (as this book does not) the IPCC's argument that added carbon dioxide could cause considerable warming, and he accepts that some curbs on carbon emissions are needed. I believe he does this to avoid being called a "denier." But, more importantly, he shows alarmists that *even using the alarmist doomsday assumptions, the Net Zero by 2050 approach makes no political, economic or social sense.*

That said, in his many articles and books Lomborg consistently tries to calm climate alarmism, and makes the same point over and over again: if we don't get too caught up in Net Zero alarmism, we will be richer in the future. If there are climate and "extreme weather" costs in the future, they will take up only a fraction of our increased future wealth. How much will we lose? How much will we gain?

Comparing 'mitigation' and 'adaptation' costs

Lomborg has extensively compared the costs of Net Zero climate *mitigation* (stopping climate change) to the expected Net Zero climate *benefits* (e.g., of the Paris Climate Agreement of 2015).

Using UN figures, he estimates that the Paris agreement, if actually honored, would cost $1-trillion in lost global GDP (best-case scenario), and more likely twice that.[18] As he points out, this is money not available for human-welfare needs, like health care, education, and so on, which are, or should be, more pressing.

However, after spending this $1- to $2-trillion (or more) a year for 80 years—for a total of $80-$160 trillion—Lomborg estimates (again, using the Paris Agreement's calculations) that by 2100 we will enjoy a temperature reduction of—drum roll please!—0.03-0.04°C. A reduction of 3-4/100th of a degree Celsius is too small to be perceived by the human senses.[19]

This is insane. Only crazy or very stupid people would pay $1- to $2-trillion a year over 80 years for this tiny a benefit.[20] In terms of a cost-benefit analysis, Lomborg expects each dollar spent on meeting the Paris Accord to produce a benefit of 11 cents.

What if we relax our efforts toward Net Zero and become less alarmist? As noted above, if we do nothing, or very little, about carbon emissions, over those 80 years humanity will gradually become much better off.

We will, therefore, be much better able to pay the costs of *adapting* to whatever "climate changes" do occur (this "warming", if the evidence in this book is correct, will be *considerably less extreme* than the IPCC and its alarmist chorus predict, and therefore much less costly than we might fear).

GDP much higher under 'business as usual'

The takeaway point in Lomborg's analysis is that under SSP5 the GDP per person will be *much higher* than today, i.e., we'll all be much better off.

Yes, we may have to spend 5.7 per cent of GDP on climate adaptation (under the alarmist assumptions of disaster ahead, which may be wrong); we could also just see ourselves as 5.7 per cent *less rich* than we would otherwise be.[21]

So the question is, can we live with this relatively small loss of a much larger income as the price of keeping the flourishing economy that is possible only (at the moment, anyway) by continuing to use fossil fuels? Lomborg believes this is an acceptable trade-off, and much better than the "burn it to the ground" economic approach of the Net Zero alarmists.

In the next chapter we'll look at how the Net Zero, "scorched-earth" approach will affect the average person in the developed world. For now, let's just recognize that the United Nations is promoting two different global futures simultaneously, much like Orwell's "doublethink," which is "the power of holding two contradictory beliefs in one's mind simultaneously, and accepting both of them."[22]

On the one hand, there is a future in which, if we mostly ignore "global warming" and keep using fossil fuels to a greater or lesser extent, we are led to catastrophe. On the other hand, there is a future in which, if we mostly ignore "global warming" and keep using fossil fuels to a greater or lesser extent, we are much better off. They can't both be right.

Honestly, you can't make this stuff up, but it's the sort of incoherence that occurs when a theory (global catastrophe due to CO2 emissions) is poorly grounded in any sort of Reality.

Notes to Chapter 13

---

[1] Malte Meinschausen, at al., "The shared socio-economic pathway (SSP) greenhouse gas concentrations and their extensions to 2100." *European Geosciences Union*, Vol. 13, Issue 8, Aug. 13, 2020. Available online.

[2] Climate Change Scenarios Primer, "Mitigation." *Sense*. Available at https://climatescenarios.org/primer/mitigation.

[3] Larry Kummer, "Manufacturing climate nightmares: Misusing science to create horrific predictions." *Fabius Maximus*, Nov. 5, 2015. Kummer's article lists a number of scientists in despair because they accept the RCP8.5 scenario as most likely.

[4] IPCC, AR6, Working Group III, Section 3.2.1.1 "Reference scenarios." Available online.

[5] Z. Hausfather, G. P. Peters, "Emissions–The 'business as usual' story is misleading." *Nature* 577, 618–620 (2020).

[6] IPCC, AR6 Synthesis report Summary for Policymakers, Section C.1, p. 25.

[7] Christensen, P., et al., "Uncertainty in forecasts of long-run economic growth." *PNAS*, May 2018, Pages 5409-5414. Cited in Bjorn Lomborg, "Welfare in the 21st century: Increasing development, reducing inequality, the impact of climate policy and the cost of climate policies." *Technological Forecasting and Social Change*, Vol. 156, July 2020, 119981. Available online.

[8] OECD, "GDP long-term forecast indicator." OECD Data. Available at https://data.oecd.org/gdp/gdp-long-term-forecast.htm. This site has a dynamic graph that shows expected global GDP growth of 2 per cent or more up to 2060. The graph shows that if GDP for the OECD nations is

US$55-trillion in 2020, GDP for the OECD in 2060 will be US$101-trillion, or double.

9 Percentage figures come from McKinsey Global Institute, "Can long-term global growth be saved?" McKinsey & Company, Jan. 1, 2015. Available online.

10 For example, an organization called Earth4All, a part of the Club of Rome, calls for a radical redistribution of income, writing: "Wealth must be re-distributed more fairly to address inequality. This will enhance societal cohesion and build trust in governments to reshape markets and invest in the future. We advocate that policies are introduced to ensure the richest 10% take less than 40% of national income by around 2030, and efforts to reduce inequality even further are implemented beyond this date." Source: *Earth4All: A Survival Guide for Humanity,* Executive Summary. In short, we will have an extreme socialist-type, or maybe even communist-type, government, where some elite—the Anointed in Thomas Sowell's terms— rather than free markets will decide how much of the "earth's resources" each of us can "share." Those of us who've read some history know how this kind of communistic, top-down "sharing" ends up—totalitarian government.

11 Leimbach, et al., cited above. This study predicts GDP growth of 2.8 per cent under the SPP5 "business as usual" scenario, only 1 per cent in the SSP3 ("regional rivalry") scenario.

12 Bjorn Lomborg, "Welfare in the 21st century." *Technological Forecasting and Social Change,* Vol. 156, July 2020, 119981. Available online. The data for the graph after 2020 comes from the Organization for Economic Development and Co-operation (OECD), 2018.

13 Wikipedia, "Shared Socioeconomic Pathways."

14 For example, Christian Bryer, et al., "New consciousness: A societal and energetic vision for rebalancing humankind within the limits of planet Earth." *Technological Forecasting and Social Change,* Vol. 114, January 2017, pp. 7-15. Available online at *Science Direct.* This study adopts the "only one world" approach, which sees technological civilization as using too many global resources to be viable.

15 Editorial, "Europe pays for Green-Energy Delusions." *Wall Street Journal,* Nov. 21, 2022 (behind a pay wall). See also John Constable, "Europe's Green Experiment a Costly Failure in Unilateral Climate Policy." *Global Warming Policy Foundation,* 2022. Available online and in PDF format. There are many articles appearing in print and online detailing the economic hardships of "green" ideology on average people in Europe.

16 Lomborg, "Welfare in the 21st century," cited above. "Section 2.2: Adaptation: Less damage from future catastrophes of coastal flooding." Available online. Lomborg gives examples of the wildly exaggerated expected costs from sea-level rise in "Section 2.2: Adaptation: less damage from future catastrophes of coastal flooding."

[17] Lomborg, "Welfare," cited above. "Section 1.4: Baseline energy use."

[18] Lomborg assumes, almost certainly correctly, that governments will not be efficient in their climate policy responses and spend more than necessary. He writes: "Politicians rarely pick the most efficient climate policies that cut CO_2 at the lowest cost. This typically doubles the cost."

[19] If you are curious about how much (or little) "global warming" would be reduced by our anti-carbon efforts, the Heritage Foundation has an online version of the IPCC's MAGICC (Model for Assessment of Greenhouse Gas Induced Climate Change) calculator (the search terms are "Heritage Foundation Calculator"). MAGICC calculates how much warming would be reduced based on X reduction of global emissions. For example, according to the IPCC's own models, a 100 per cent emission-reduction by 2050 (impossible) would reduce "global warming" by a minuscule 0.13°C.

[20] Lomborg, "Welfare," cited above, "Section 7.3: Costs and benefits: Paris agreement."

[21] A U.S. government report on the costs of climate policy, released in March 2023, inadvertently makes the same point (inadvertently because the report doesn't predict climate catastrophe as the Joe Biden government does). Looking at 12 scientific studies on how much U.S. GDP would decline as temperatures increase, the report found that on average the increased "global warming" of 1.2°C (2.2°F) since 1851-1900 reduced GDP by 0.5% (half a percentage point). However, since 1950, GDP had risen by 800 per cent, so this 0.5% in lost GDP is a trivial amount and hardly constitutes a "climate emergency," at least as far as the economy is concerned. Projecting ahead, and again taking the average of the 12 studies, the report suggests that if temperatures went up to 2.5°C (4.5°F) of pre-industrial levels ("catastrophe" territory!), the U.S. GDP would fall by 2 per cent. If GDP increased by a rather low 1.5 per cent a year over the next 80 years, the U.S. economy would still grow by 232 per cent. Subtracting the 2 per cent "climate change" penalty means GDP growth would be "only" 225 per cent—7 per cent less over 80 years. This is, in economic terms, not significant and means "climate change" will not be catastrophic for the U.S. economy. And if the economy can thrive, why can't its people? Source: Council of Economic Advisors & Office of Management and Budget, "Methodologies and Considerations for Integrating the Physical and Transition Risks of Climate Change into Macroeconomic Forecasting for the President's Budget." March 13, 2023. The analysis of the report data comes from Steven E. Koonin, "The White House tells the truth on climate." *Wall Street Journal*, July 7, 2023.

[22] George Orwell, *1984*. Toronto: Penguin, 2008 (1949), p. 37.

Sidebar 6: Radiative forcing

Radiative forcing is the amount of warming produced by a given amount of carbon dioxide, water vapor, or any other greenhouse gas, expressed as Watts per square metre (W/m^2).

Radiative forcing is discussed in the IPCC's sixth report (AR6) but, inexplicably, in the attempt to persuade us that we face dangerous warming, the IPCC gets the physics wrong.

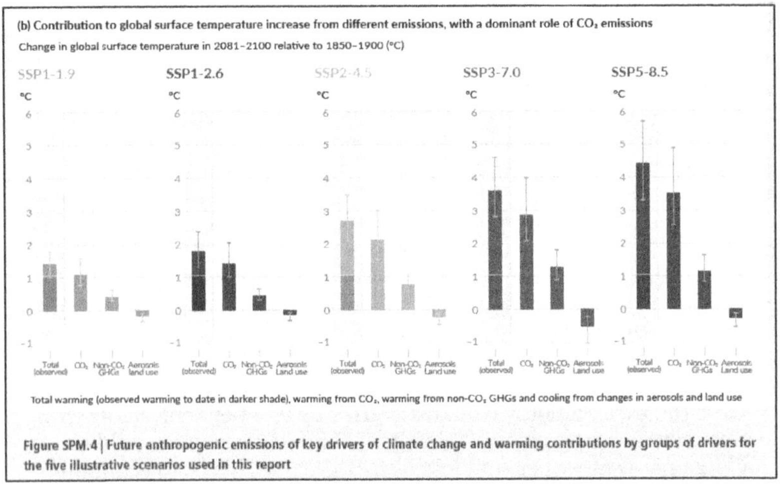

Figure SB6.1: Radiative forcing SSP scenarios. Source: IPCC AR6 Summary for Policymakers, Section B, Box SMP1, p. 13

Figure SB6.1, from the AR6 Summary for Policymakers, shows the amount of warming we can expect for each of the Shared Socioeconomic Pathway (SSP) scenarios in the sixth IPPC report, based on how much radiative forcing they will produce (again, the number after the hyphen for each SSP is the amount of radiative forcing for each scenario in Watts per square metre).

So, for example, SSP5-8.5, on the right, predicts a "Total (observed)" of 4.5°C of warming if radiative forcing reaches 8.5 W/m^2. Scary! SSP1-1.9, on the left, predicts 1.5°C of warming from radiative forcing of 1.9 W/m^2. And so on.

The problem is that in science that is "normal," rather than "post-normal," and that therefore actually pays attention to the laws of physics, there is something called the Stefan-Boltzmann Law that measures the amount of heating you *actually* get for various heat increases in Watts per square metre.

The discussion below comes from a Tom Nelson podcast interview with physicist Howard Hayden, available on YouTube.[1] I had to watch it three times to understand what Hayden was saying and I'm summarizing a portion of it for you below. I'd recommend watching the whole video as a pointer to how the IPCC tends to bend "The Science" in pursuit of its ideological goals.

By the way, the formula for the Stefan-Boltzman Law is complicated, but fortunately a gadget on the website Omnicalculator[2] will calculate the amount of heating you get from a given W/m^2 increase due to greenhouse gases (see Figure SB6.2).

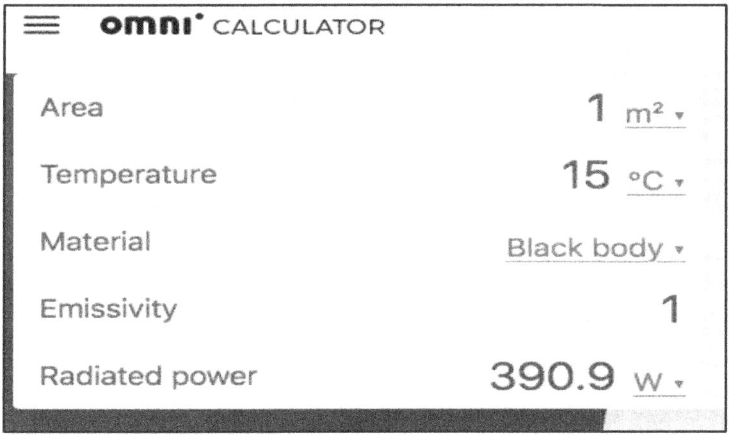

Figure SB6.2: Stefan-Boltzmann calculator. Source: omnicalculator website

As you can see, if we input an area of 1 square metre, set the "Material" to "Black body", and provide a base temperature of 15°C (the estimated average global temperature[3]), we get a radiative forcing of 390.9 W/m^2. That is, it took 390.9 W/m^2 of radiative forcing to generate 15°C of warming.

If we use the Omnicalculator to increase the temperature by 1°C, to 16°C, the total radiative forcing becomes 396.4 W/m^2, so an increase of 5.5 W/m^2 produced this extra 1°C of warming.

If the temperature goes up another 1°C, to 17°C, the total radiative forcing would be 401.9 W/m^2, an increase in radiative forcing of 5.5 W/m^2 from 16°C and therefore about 11 W/m^2 above the 15°C temperature. On average, then, an increase of 5.5 Watts per square metre creates a warming increase of 1°C.

However, if we zero in on SSP5-8.5 (the "business-as-usual", fuel-guzzling scenario) in Figure SB6.3, and remembering that the SSP number after the hyphen equals the amount of additional radiative forcing, you will see that for the IPCC, radiative forcing of 8.5 produces warming of 4.5°C, with 3.5°C of this warming coming from CO2.

However, if we apply the Stefan-Boltzmann Law (and, as an aside, this is a "law" of physics, not the result of a "consensus" that we can accept or reject as the "post-normal" whim takes us), 8.5 W/m^2 can only give us about 1.5°C of warming.

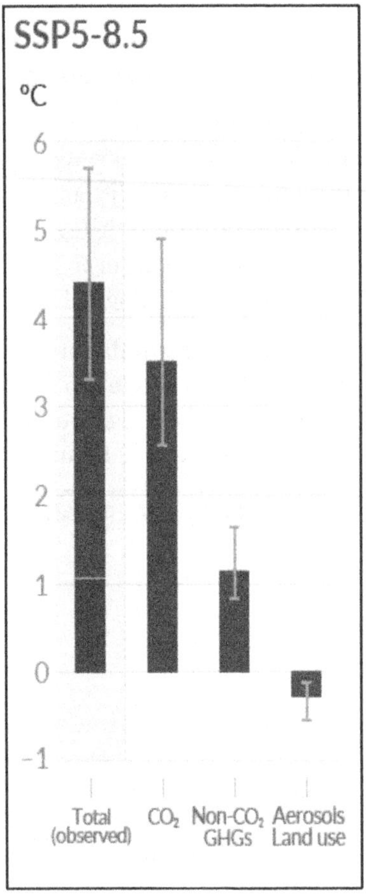

SB6.3: SSP5-8.5 close-up

That is, as we saw above using the calculator, if 5.5 W/m² produces 1°C of warming, and 11 W/m² produces 2°C of warming, then 8.5 W/m² can only provide warming of about 1.5°C.

In the same way, at least if we stay inside the bounds of Real World physics, carbon dioxide alone (the second vertical bar in the figure) cannot produce 3.5°C of warming from radiative forcing of 8.5W/m².

If the Stefan-Boltzmann Law is correct, which most traditional (that is, not "post-normal") physicists accept, then to get 4.5°C of warming we would have to apply radiative forcing of about 25 W/m², not 8.5 W/m².

And the same is true for the other SSP numbers in Figure SB8.1, as you'll see if you use the calculator and do the math. The SSP radiative-forcing numbers for each SSP, expressed in W/m², *are not large enough to produce the warming claimed.* So either the warming is being magnified in some way that the IPCC doesn't care to explain, or these scary warming numbers (like 4.5°C) have just been pulled out of thin air.

On the surface of it, at least, it appears the IPCC, or its writers of the Summary for Policymakers, are once again trying to deceive and frighten us by using figures that, at least in the Real World (as opposed to the "post-normal" Wonderland world), don't match reality.

Notes to Sidebar 6: Radiative forcing

---

[1] "#53: Howard Hayden: The IPCC does *not* apply the Stefan-Boltzmann law to their results." *Tom Nelson Podcast*, 2022. The URL is https://youtu.be/LPPoItFaees.

[2] The URL for the Stefan-Boltzmann Law calculator is https://www.omnicalculator.com. You will get the Stefan-Boltzmann Law if you put this term in the search window. This website has a dizzying array of equations from many scientific disciplines and is really cool!

[3] Rebecca Lindsay & Luann Dahlman, "Climate change: Global temperature." *Climate.gov*, Jan. 18, 2023.

Chapter 14: How Net Zero will make us poorer—a lot poorer

"You couldn't have it if you did want it," the Queen said. "The rule is, jam tomorrow and jam yesterday—but never jam today." "It must come sometimes to 'jam to-day'," Alice objected. "No, it can't," said the Queen. "It's jam every other day: today isn't any other day, you know."

We're assured by our governments and the IPCC that, once this messy business of getting the "climate crisis" under control is over, we'll all be better off and even happier, since our lifestyles will be much more "sustainable."

Indeed, as a video that appeared on the World Economic Forum website promised, "You will own nothing and you will love it!" The video was quickly removed as rather too revealing of the WEF's ultimate agenda for us all: less of everything. But less-is-better-and-you-will-love-it is the direction current climate policy is taking us.

This beneficent vision is the essence of the IPCC's SSP1 scenario: happy, fully co-operating, non-competitive beings in an environmental and economic Garden of Eden. SSP1 is described this way:

> Shared Socioeconomic Pathway 1: Sustainability – Taking the Green Road (Low challenges to mitigation and adaptation)
>
> The world shifts gradually, but pervasively, toward a more sustainable path, emphasizing more inclusive development that respects perceived environmental boundaries. Management of the global commons slowly improves, educational and health investments accelerate the demographic transition, and the emphasis on economic growth shifts toward a broader emphasis on human well-being.

> Driven by an increasing commitment to achieving development goals, inequality is reduced both across and within countries. *Consumption is oriented toward low material growth and lower resource and energy intensity.*[1] [emphasis added]

This sounds amazing!

Unfortunately, humans have almost never achieved this level of social, economic and environmental harmony before, at least not on a large scale.

And humans are rarely happy with low material growth and *fewer* resources and energy. This is why attempts to reach this kind of beatific, low-consumption future always require force in the end (and a fully implemented Net Zero will need lots of force) and have almost always descended into tyranny. But this time will be different! Of course it will.

But getting back to reality, let's look at what the various Net Zero fantasies will *actually* cost us, the average person who doesn't belong to some economically privileged and protected elite (we're looking at you, Leonardo di Caprio). Under Net Zero you will own a lot less (a.k.a., be poorer), but will you "love" your more impoverished lifestyle? Will you be, in fact, "happier"?

We saw in Chapter 13 that Bjorn Lomborg's estimated cost of $1-$2-trillion a year for the Paris Agreement will reduce "global warming" by effectively zero (about 3- to 4/100ths of a degree Centigrade).

That's because, as Lomborg notes in another calculation, the Paris Agreement, if fully observed by all the signatory nations (unlikely), would only remove 63 Gigatons of carbon from the atmosphere, which (using IPCC estimates) is enough to reducing warming by—wait for it!—0.029°C (in effect, 3/100ths of a degree Celsius).

Obviously, these figures for warming reduction are both fluid and largely imaginary, since we don't have "facts" from the future. We can, however, be sure that for all our efforts the amount of "global cooling" will be minuscule.

To keep the warming increase to only 1.5°C, Lomborg notes, the world would have to reduce carbon emissions by *6,410 Gigatons* (again, Lomborg is using the IPCC-model assumptions that temperature is highly "sensitive" to changes in CO_2 levels which, as we've seen, is not the case).

In other words, the emission reductions promised by the Paris Agreement are *1/100th* of what is needed to achieve the target 1.5°C.[2] Which means Paris is a delusion, but a very expensive one.

Cost of Net Zero for Canadians

Just for fun, let's see what Net Zero by 2050 might cost the average Canadian family. The Royal Bank of Canada, which has adopted an alarmist corporate position, estimates the fight to reduce "climate change" will cost Canadians $2-trillion over the next 30 years, or about $60 billion a year; this, the bank says, would cut Canada's emissions by 75 per cent from current levels.[3]

RBC considers this expense "affordable," in part because we'll be inventing amazing new "green" technologies and, as part of the so-called "just transition," creating thousands of new "green" jobs to replace the nasty (but generally high-paying) fossil-fuel-based jobs.

As a point of comparison: Canada's entire GDP was less than $2-trillion in 2022.[4] As another point of comparison, Canada's federal deficit for 2023 is expected to be a record $40-billion. And we're going to add $60-billion to that? Or make cuts worth $60-billion each year? Is this even feasible?

A report by McKinsey & Company, a consulting group that fully supports global-warming alarmism, estimates that a full-bore transition of physical and electrical assets to Net Zero by 2050, with the aim of keeping warming to 1.5°C, would cost $275-trillion globally, or about $9.2-trillion a year.[5]

Like RBC, the McKinsey report also lauds the many "gains" and "opportunities" it sees in the transition to a "green" economy, but it's hard to believe these economic "advantages" will be anywhere near $9.2 trillion a year, even by 2050.

On a cheerier note, the report tells us that this $9.2-trillion a year is only (!) $3.5-trillion more than we, globally, are currently spending on anti-carbon measures.

That's a relief! We're currently only spending $5.7-trillion a year on climate measures that will, if Lomborg is correct, have no measurable impact on temperatures. Meanwhile, at least in B.C., a million people don't have a family doctor, home-ownership is out of reach for most young people, and people are living in tents on the streets because there aren't enough addiction-treatment facilities.

Which is just to say that there are many other, far more useful ways that we could be spending this $5.7-trillion (or $9.2-trillion). For example, as Bjorn Lomborg has suggested many times, rather than trying to make fossil fuels more expensive to discourage their use, spend the Net Zero money on R&D to develop "green" technologies that aren't ridiculously expensive and that people might actually *want* to buy.

On a more ominous note, McKinsey estimates that if we don't take measures against "climate change," if we just blithely continue with "business as usual," the cost would be $250-trillion in climate damage. So, doing the math ($275-trillion minus $250-trillion = $25-trillion), we're back to Lomborg's Paris Accord estimate of $1-$2-trillion a year up to 2050 to (maybe) keep the temperature increase to 1.5°C.[6]

That said, while $275-trillion over 30 years sounds about right for a full-on attack on "global warming," $250-trillion ($8-trillion a year up to 2050) in climate damage seems absurdly high given that, as Lomborg suggests, we won't be sitting idly by while the climate throws challenges at us. We'll respond to reduce the damage.

Vaclav Smil: What $275-trillion looks like

As noted above, keeping the global temperature increase to 1.5°C could cost $275-trillion by 2050. Terms like "billions" and "trillions" are thrown around so freely today that they have almost no meaning for most of us. So Elon Musk spent $44-billion to buy Twitter (renamed X)? Ho hum. An extra $3.5-trillion a year, as McKinsey suggests? Chump change!

We have this cavalier attitude because most of us (I include myself) haven't a clue what a "trillion," much less a "billion," really looks like. They're just *words*; they don't affect us. Or do they? (I'd say the same is true of terms like "climate change"—they are so vague as to be meaningless. But I digress....)

Vaclav Smil is a Canadian environmental scientist with an extensive body of published articles and books. Figure 14.1 is his illustration of what $275-trillion dollars actually looks like (each grey cube equals US$1-billion). He also compares this spending to the cost of previous

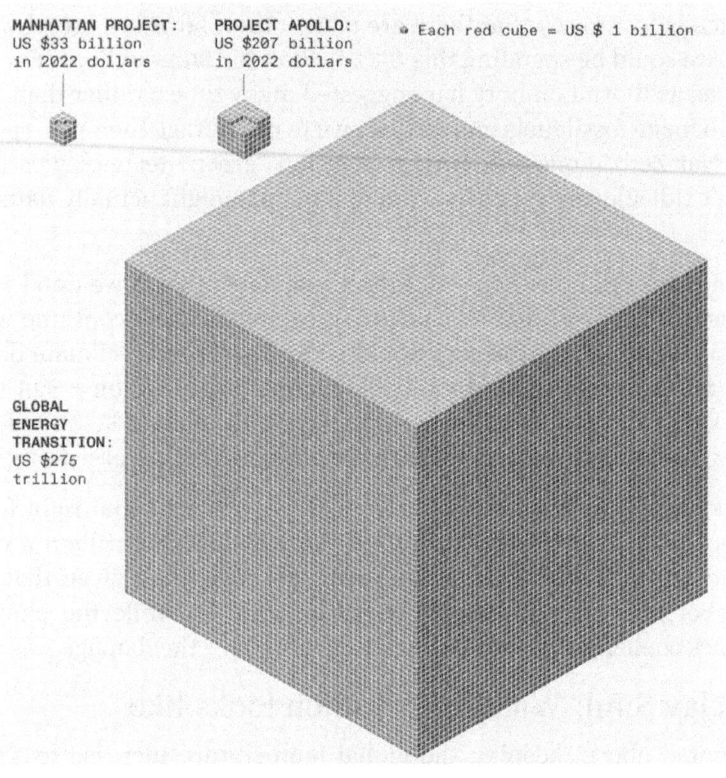

Figure 14.1: Comparing the cost of decarbonization with the costs of the Apollo Project and Manhattan Project. Each red cube equals US$1-billion. Source: Chris Philpot; CTBTO Preparatory Commission; Science Direct: McKinsey Global Institute

mega-projects—the Manhattan Project (the first atomic bombs) and the Apollo Project that put humans on the moon.[7]

Smil notes that the Manhattan Project cost $2-billion, or $33-billion in 2022 dollars. The Apollo Moon Project cost almost ten times more: $26-billion, or $207-billion in today's money. The Manhattan Project cost about 0.3 per cent of the U.S.'s GDP from 1943-45; Apollo cost 0.2 per cent of the U.S. GDP from 1961-72. By comparison, the McKinsey report predicts, decarbonizing the world economy will cost about 10 per cent of global GDP.

Smil's point is that the two previous mega-projects were both accomplished with a fraction of a percentage of the U.S. GDP of the time and didn't affect the economic lives of the average American that much. At a cost of ten per cent of GDP, on the other hand, the decarbonization of the global energy supply will impact everyone, and not in a positive way. Smil writes:

> Ours is an overwhelmingly fossil-fueled civilization, and the size and complexity of our extensive supersystem of fuel extraction, processing, distribution, storage, and conversion means that a complete displacement of it will directly affect every person and every industry, not least the growing of food and the long-distance transport of goods and people. The costs will be stupendous.[8]

As for the percentage of GDP lost to this massive climate-mitigation project, Smil notes:

> Because the world's low-income countries could not carry such burdens, affluent nations would have to devote on the order of 15 to 20 percent of their annual economic product to the task. Such shares are comparable only to the spending that was required to win World War II.[9]

Would we, average citizens, be prepared to take this massive 15-20 per cent (or perhaps more!) hit to our standard of living to fight "global warming" *if we were told in advance what these costs would be*? Probably not, which is why our various governments are cagey about presenting us with the actual bill.

An internet search with the terms "What will Net Zero cost Canadians?" or similar yields remarkably few hits that actually *quantify* what the Canadian government expects the cost achieving a Net Zero society will be.

Governments cagey on costs of Net Zero

This lack of transparency is nothing new. Way back in 1999, then-*Globe and Mail* science reporter Lydia Dotto jumped on the global-warming bandwagon early with her book *Storm Warning: Gambling with the Climate of Our Planet.* In her discussion of what the 1997 Kyoto Protocol would cost, she notes that the expected costs were never spelled out to the Canadian public.[10]

She quotes from a 1997 column by then-*Globe and Mail* columnist Jeffrey Simpson (another climate alarmist author[11]) in which Simpson writes:

> Only when people appreciate how precisely their lives will change, what those changes will cost and when the changes must be implemented, can a reasonable measure be taken of public opinion. *For now, people have only the haziest idea, if they have any at all, about how their lives must change.*
>
> And change they must. Emissions must decline by about one-fifth in the period 2000 to 2010 for Canada to meet its Kyoto target. No reduction of that incredible magnitude can be achieved without wrenching changes. *At no time has the [Jean] Chrétien [Liberal] government spoken truthfully to Canadians about what lies ahead.*[12] [emphasis added]

I repeat: Simpson was totally on-board with the alarmist program. He's just a little more candid than our governments that there are "wrenching changes" ahead. "Wrenching changes!" What does that mean for the average household, keeping in mind that the "wrenching changes" of Kyoto were a walk in the park compared to what the Net Zero promoters plan for us?

In 2007, Environment Canada estimated that reducing CO2 emissions by 6 per cent below 1990 levels, as sought by the then-Stephane "Green Shift" Dion-led Liberals, would cost 275,000 jobs, increase electricity prices 50 per cent, increase gasoline prices by 60 per cent, and cost each Canadian family $4,000 a year in lost income.[13]

Another Environment Canada report the same year offered the following:

> The Government's analysis, broadly endorsed by some of Canada's leading economists, indicates that Canadian Gross Domestic Product (GDP) *would decline by more than 6.5% relative to current projections in 2008 as a result of strict adherence to the Kyoto Protocol's emission-reduction target for Canada.*
>
> This would imply a deep recession in 2008, with a one-year net loss of national economic activity in the range of $51-billion relative to 2007 levels. By way of comparison, the most severe recession in the post-World War II period for Canada, as measured by the fall in real GDP, was in 1981-1982. Real GDP fell 4.9% between the second quarter of 1981 and the fourth quarter of 1982.[14] [emphasis added]

Moving to more recent times, a 2023 column by Campbell Clark in *The Globe and Mail* observes that we *still* aren't being told by politicians what the costs of fighting "climate change" will be. He writes:

> Say, politicians, is there a cost to your climate-change policies? The answer is yes. Don't try to avoid that fact. You can't. The costs are coming one way or another. The real question is whether those costs bring benefits. *The failure to acknowledge the cost side of the calculation amounts to fantasy thinking*, and it's one of the reasons that Environment Minister Steven Guilbeault, a lifelong environmental activist, fails miserably every time climate-policy debates are translated into dollars.[15] [emphasis added]

In other words, the Canadian government (and other governments) have known for decades that these climate-change policies would be ruinously expensive, both for government and the average citizen.

This is why governments, wisely, have decided it's best not to tell us what we will be expected to pay to "fight climate change" today.

The politicians have also, in almost all cases—and perhaps moved by a knowledge that these anti-carbon measures if fully implemented will ruin their economies and election chances—only moved partially toward their Kyoto and Paris commitments.

If we *were* told, honestly, the cost of "fighting climate change," I have no doubt we would say "No, thanks, we'll take our chances with 'global warming'." But perhaps that's just me...

Net Zero costs 'only' half of global corporate profits

Getting back to the McKinsey Report and it's $275-trillion estimate for a successful Net Zero:

To ease any anxieties we Canadian taxpayers might have about this increased level of spending, the report helpfully observes that the *extra* $3.5-trillion a year (added to the $5.7-trillion we are already spending on climate mitigation for a total of $9.2-trillion a year, as noted above) is equivalent to a mere *half of global corporate profits, one-quarter of global total tax revenue, or 7 per cent of global household spending.*[16]

This level of expenditure begs the questions (among others):

- Will corporations willingly sacrifice half their earnings to the chimera of fighting "climate change"? [Rough guess? No.]
- Are governments willing to give up one-quarter of their total tax revenues? [Rough guess? No.]
- Are you, personally, willing to sacrifice seven per cent of your household spending every year? [Rough guess: probably not.]

Given what I think are the most likely answers to these questions, maybe an extra $3.5-trillion a year isn't a trifle after all....

Diving a bit deeper: The global GDP in 2022 was about $104-trillion,[17] so, again, if McKinsey's report is correct, we're talking about spending almost 10 per cent of global GDP (that minuscule $9.2-trillion) each year to keep "climate change" to 1.5°C.

It's hard to believe that the governments of the world (apart, perhaps, from Canada under the ultra-green Justin Trudeau Liberals) would be willing to reduce their economic growth by 10 per cent each year to keep the temperature increase to 1.5°C when, again, they can't even meet their full Paris Agreement promises of $1-$2 trillion a year, not to mention that there's no guarantee that the climate will cool appreciably, or at all, if we did go all in.

Net Zero cost to households—it won't be zero

In household terms, imagine that you will lose seven per cent of your spending money each year for the next 30 years to fight "climate change."

The average Canadian household income is estimated at $96,000 per year as of 2020[18]; under the Net Zero by 2050 plan, then, each household will, on average, be contributing $6,700 each year to the "climate change" fight, and this is *in addition to* whatever the household is already paying in taxes.

Obviously, this isn't what each *actual* Canadian household will pay—presumably better-off Canadians will be paying more, the less-well-off will be paying less.

And, since this is a global figure, as Vaclav Smil notes above, the expense will fall harder on the citizens of the developed nations than those of the less-developed nations, so Canadians will actually be paying more than seven per cent per household per year, and probably a lot more. Will it be $10,000 a year? $20,000 a year? $30,000 a year?

Perhaps average Canadians (like Americans, see below) think the cost will be borne by government and taxes on the big corporations, not them. Unfortunately, there's only one taxpayer.

Even if the various national governments borrow to pay for this extra expense, taxpayers will still be on the hook each year for the interest on the debt, which increases as the national debt increases. (It's Alice in Wonderland thinking that, as Justin Trudeau said while campaigning in 2014, "the budget will balance itself."[19])

And if governments increase taxes on corporations, then the goods and services that companies sell will be more expensive (Economics 101).

However this additional expense is financed, we will overall, as households and individuals, have less money to spend (a.k.a., will be "poorer") as our contribution to Net Zero.

How much are we willing to spend to stop 'climate change'? Not that much....

Perhaps most of us are happy to be poorer if it saves us from "global warming." Alas, studies on how much the average person, if given the choice, is *actually* willing to pay in the "existential" battle against "climate change" say the opposite.

For example, a *Washington Post* poll found that 76 per cent of Americans regard "climate change" as "a crisis" or "a major problem but not a crisis." So far, so good.

But fewer than 40 per cent expect to have to make a "major sacrifice" to fight climate change. Fewer than half were willing to pay an extra $2 a month on their electricity bill (a whopping $24 a year); a quarter said they'd be willing to shell out—wait for it!—$10 a month.[20]

The Americans surveyed want more fuel-efficient vehicles, but most are unwilling to accept an increase in the gasoline tax.

And, as noted above, most expect corporations and government to shoulder the expense of fighting climate change, as if corporations and governments live in another universe than consumers and taxpayers. Given the *multi-trillions* of dollars being bandied about as the cure for climate woes, these Americans (and by extension, Canadians) are living in an Alice in Wonderland fantasy world.

Nor are Europeans any more realistic than Americans about the financial impact of Net Zero. A survey of people in ten countries, including the U.S., Britain, France and Germany, found 78 per cent of respondents were "personally concerned about climate change," and 76 per cent wanted tighter environmental regulations. Score a point for the IPCC!

But wait—almost *half* saw no need to change their personal habits. They felt they were doing enough through, e.g., recycling plastics, buying low-power lightbulbs, etc. In other words, most had fine intentions, as long as they don't have to actually spend very much of their own money.[21]

When the economic Net Zero freight train hits these well-meaning people, it's going to be quite an eye-opener! Especially since most of us seem to be supporting Net Zero with our eyes closed to the possible financial and social impacts.

Cities oblivious to impacts of Net Zero

As an example of Canadians' incomprehension of the potential impact of Net Zero policies, in 2022 Ottawa's city council fought, successfully, against an upgrade of the natural-gas pipeline that supplies the city.

The pipeline owners, Enbridge Gas, argued that the pipeline could suffer a "catastrophic failure" in the next few years if not upgraded. For city council, however, reliance on natural gas wasn't "green" and violated its official "Energy Evolution" plan, which increasingly relies on renewable energy.[22]

This means Ottawa residents are counting on "renewable energies" to keep them warm in winter (reality check: despite "global warming," Ottawa still has *very cold* winters) if the natural-gas pipeline fails, even though renewables are nowhere near ready to bear this load. Fingers crossed, then.... This decision by Ottawa council is, by the way, completely irrational, sacrificing energy resilience for green ideology.

One more example of the delusional approach a U.S. city (it's not just Canadian cities going insane) is adopting in pursuit of green dreams (I could cite many more).

The City of New York wants to reduce carbon emissions by 40 per cent by 2030. In practical terms, this means the city's electrical grid will be free of fossil fuels (mostly natural gas) by that date, relying more and more on "renewables" like solar and wind power.

Unfortunately, solar and wind aren't 24/7 reliable, so the city will need battery backup from time to time since fossil-fuel power plants will no longer be available and New York State has closed its nuclear power plants. In other words, there is no conventional power backup if the sun doesn't shine or the wind doesn't blow. That means storage batteries.

So how much backup battery storage power would be needed to cover the renewables' "down time" for the city? Roughly 10,000 Gigawatt hours of battery storage (a Gigawatt is one billion watts).[23] Using the cost of a Tesla battery as a guideline, the author I'm citing here estimates this amount of battery power would cost the city about $1.5-trillion, which is roughly the entire yearly GDP of New York State.[24]

In short, New York City, like the City of Ottawa, has set itself an impossible—a.k.a., "insane"—goal of near-total "greenness." If New York doesn't invest the $1.5 trillion on batteries (and it won't because because the cost is unaffordable), then when these "green" energy sources falter and fail, as they will from time to time, millions of people will pay the price; in winter, this failure could mean their very lives.

I hope I've given you a glimmer of an idea—and it's just a glimmer; the reality may be much worse than we can imagine—of what awaits us, financially, if we embrace the Net Zero policies: We will be *much poorer* while, in terms of climate, we might be a fraction of a degree Celsius cooler—and that's assuming temperature is as "sensitive" to CO2 as the IPCC believes, which it isn't. In fact, our Net Zero efforts will almost certainly give us "zero" results in terms of cooling, or very close to zero, at enormous cost.

For any rational person, Net Zero is an insane proposition. But the situation gets worse when we consider the effect of Net Zero policies on our individual and social freedom. We'll look at this issue in the next chapter.

Notes to Chapter 14

---

[1] This scenario description comes from Keywan Riahi, et al., "The Shared Socioeconomic Pathways and their energy, land use, and greenhouse gas emissions implications: An Overview." *Global Environmental Change*, Vol. 42, January 2017, pp. 153-168. Available online at *Science Direct*.

[2] Bjorn Lomborg, "Section 4.2: Benefits of the Paris Agreement." Welfare in the 21st Century. *Technological Forecasting and Social Change*, Vol. 156, July 2020.

[3] Royal Bank Special Reports, "The $2 trillion transition: Canada's road to Net Zero." October 20, 2021. Available online.

[4] Geoffrey Morgan, "Canada's road to Net Zero will cost $2 trillion—as much as a year's GDP." *Financial Post*, Oct. 20, 2021. Available online.

[5] McKinsey Global Institute, *The Net Zero Transition: What it could cost—what it could bring*. McKinsey & Company, January 2022, "In brief," p. viii. Available online and in PDF format.

[6] McKinsey, *Net Zero Transition*, p. 19.

[7] Vaclav Smil, "Taming the Climate is Far Harder Than Getting People on the Moon: Decarbonization is a project with no clear beginning or end." *IEEE Spectrum*, Sept. 29, 2022. Available online.

[8] Smil, *Ibid.*

[9] Smil, *Ibid.*

[10] Lydia Dotto, *Storm Warning: Gambling with the Climate of Our Planet*. Toroto: Doubleday Canada, 1999, p. 280.

[11] Simpson is the co-author, along with climate scientist Marc Jaccard and researcher Nic Rivers, of *Hot Air: Meeting Canada's Climate Challenge*, Douglas Gibson books, 2007.

[12] Jeffrey Simpson, "Ottawa blows a lot of hot air about greenhouse gases." *Globe and Mail*, Dec. 18, 1997.

[13] Environment Canada, "The cost of Bill C-288 to Canadian families and business." April 2007. Available online.

[14] Environment Canada, "A climate change plan for the purposes of the Kyoto Protocol Implementation Act—2007." Sept. 6, 2007. Available online.

[15] Campbell Clark, "There is no cost-free climate plan." *Globe and Mail,* May 22, 2023.

[16] McKinsey Report, p. viii.

[17] Avery Koop, "The $100-trillion global economy in one chart." *Visual Capitalist,* July 12, 2022. Available online.

[18] Statista Research Department, "Median annual family income in Canada in 2020, by province." *Statista.* Available online. See also, Gregory Rozdeba, "Average income in Canada (2023)." *Dundas Life,* April 17, 2023. Available online. The Dundas study puts the average Canadian salary at $59,000; the total of $96,000 for a two-income family seems reasonable.

[19] "Editorial: Justin Trudeau's budget still hasn't balanced itself." *London Free Press,* March 1, 2018.

[20] Brady Dennis et al., "Americans Increasingly See Climate Change as a Crisis, Poll Shows." *Washington Post,* Sept. 13, 2019. Cited in Bjorn Lomborg, *False Alarm: How Climate Change Panic Costs Us Trillions, Hurts the Poor, and Fails to Fix the Planet."* New York: Basic Books, 2020, p. 227.

[21] Jon Henley, "Few willing to change lifestyle to save the planet, climate survey finds." *The Guardian,* Nov. 7, 2021. Available online.

[22] Patricia Adams & Lawrence Solomon, "Ottawa is committing suicide." *Probe International,* June 24, 2022. See also Mitchell Beer, "Ontario Regulator Refuses New Pipeline, Tells Enbridge to Plan for Lower Gas Demand." *The Energy Mix,* May 29, 2022.

[23] Francis Menton, "My Testimony on New York's 'Scoping Plan' to Achieve Net Zero Carbon Emissions." *Manhattan Contrarian,* May 3, 2022. Available at https://www.manhattancontrarian.com/blog/2022-5-3-my-testimony-on-new-yorks-scoping-plan-to-achieve-net-zero-carbon-emissions.

[24] I did my own calculation, based on the 10,000-Gigawatt figure. The battery for a low-end Tesla costs $13,000 (minimum) and generates 50 kilowatts of power (a kilowatt equals 1,000 Watts). The cost per watt is 26 cents. The cost, then, of 10,000 Gigawatts of Tesla battery power would be $2.6-trillion. Francis Menton's figure is lower because, obviously, a city like New York would be buying battery power in bulk, as it were, and get a cheaper price. One hopes.

Chapter 15: Learning to love poverty

It's a great huge game of chess that's being played—all over the world—if this is the world at all, you know.

A British universities research group called UK Fires investigates ways and means to reach zero carbon in that country by 2050. Specifically, it lists what the average Briton needs to do—or more accurately, not do—to help that nation reach its Net Zero target.

These "do's" and "don't-do's" include:

- stop using airplanes, and lobby for the almost total elimination of airports;
- only heat rooms you are in (Brits are already starting to do this involuntarily thanks to massively increased heating costs, just as they did in the decades after World War II);
- avoid using cement as a building material and rely on recycled materials only;
- eat less meat;
- and much more.[1]

While UK Fires is on board with Net Zero, its reports make it very clear why the rest of us *should not be on board* unless we are happy to accept a disastrous fall in our standard of living and what amounts of a global economic collapse.

In his book *Heat: How to Stop the Planet from Burning*, ultra-alarmist George Monbiot quotes a veteran British environmentalist who, at a meeting, was asked: "What would Great Britain be like if it cut carbon use by the 80 per cent needed to prevent runaway global warming?" The environmentalist's reply? "A very poor third-world country."[2]

A "very poor third-world country"—is that what most of us want? In fact, this sounds like a bad joke. It's no joke. Alarmists like Monbiot and many others are dead serious: they want to *substantially* reduce our standard of living on the theory (so far unproven in the Real World) that the "planet" can't "sustainably" support eight billion or more of us at a Western standard of living.

'Making Peace with Nature'

For example, a 2021 United Nations' Environmental Program (UNEP) report entitled "Making Peace with Nature" notes that environmental degradation has gotten so bad that

> only a system-wide transformation will achieve well-being for all within the Earth's capacity to support life, provide resources and absorb waste. This transformation will involve a fundamental change in the technological, economic and social organization of society, *including world views, norms, values and governance.*[3] [emphasis added]

The report continues:

> With successful transformative change, the consumption of resources would decrease in wealthy contexts and increase sustainably elsewhere. ... Human ambitions for a good life *would no longer be centred around high levels of material consumption, but around rich relationships involving people and nature,* in keeping with diverse traditions throughout the world.[4] [emphasis added]

Sounds great! A change of "world views, norms, values and governance"—no big deal.

But what about those stubborn and unevolved people who rather like "high levels of material consumption" *as well as* "rich relationships involving people and nature"? Alas, there will be holdouts, the UNEP report tells us:

> Opposition from vested interests to transformations aiming to secure a sustainable and prosperous future is to be expected but can be addressed. Existing infrastructure and built capital provide system inertia that can make change

> difficult and incurs short-term costs, especially if change involves the premature retirement of capital stock.
>
> Also, individuals and organizations have habits, procedures and ways of doing business that can yield a reluctance and resistance to change. Individuals and organizations can also oppose change that disrupts their livelihoods, market share and revenues, or that otherwise appear unfair.[5] [emphasis added]

Imagine! Individuals and organizations with the gall to oppose measures that will disrupt their "livelihoods, market share and revenues" or that—gasp!—"appear unfair"! Fortunately, the report has an answer for dealing with these economic and social Luddites:

> Some opposition can be addressed proactively by redirecting subsidies to steer workers and firms toward opportunities associated with transformative change. Programs fostering a just transition can include, for instance, retraining workers from unsustainable industries and helping them relocate in order to take up new jobs.[6]

Well, that's a relief. The brutish, selfish, capitalist die-hards will be gently persuaded that losing their livelihoods, etc., is a Good Thing (if history is any guide, this "proactive" persuasion will include repression, concentration camps/gulags, and maybe even executions).

On the plus side, once this ecological and social Utopia is in place, we'll all be New Men and Women, freed from our selfish human nature, and all operating on the principle of "From each according to his ability, to each according to his needs."

Although, this sounds strangely familiar, doesn't it? Where have we heard this phrasing before? Oh, yes, it's a basic plank of the communist platform.

Actually, when you think about it, isn't this UNEP report proposing something awfully like, well, communism? And we all know, or should know, how well communism has worked in the past as it tries to create a New Humanity (looking at you, Soviet Russia and Communist China, but also North Korea and Pol Pot's Cambodia, among other murderous regimes).

The environment is improving, not degrading

Also, globally the environment is actually *improving*, not degrading, at least in the developed world and increasingly in the developing world as well. This environmental improvement is, ironically, due in large part to affluence.

Yes, affluence. The thing is, as noted in Chapter 12, affluent people are willing and able to pay for a clean environment; it's a benefit of "post-materialism", as Michael Shellenberger and Ted Nordhaus put it in their book *Break Through: From the Death of Environmentalism to the Politics of Possibility*.[7]

The affluent people who write a report like "Making Peace with Nature" are a good example of Shellenberger's and Nordhaus's point. The report writers have never had to deal with nature in the raw, except perhaps through camping, so they can afford to idealize Nature. Poor people are working too hard to indulge dreams of making this kind of non-materialist "peace" with nature.

But let's say folks in the developed nations are willing to forego many of their material comforts so that those in the developing nations can have more. In the Real World, this belief is delusional.

If the capitalist global economy collapses under the "Making Peace" plan, as would happen—indeed, this collapse is considered desirable by the UN—what will *actually* happen in the less-developed world?

For example, what would be the fate of the tourist industry in, say, Mexico if almost all passenger planes are banned, as UK Fire recommends and "Making Peace" strongly implies? I can only imagine that for the five million Mexicans in the tourist industry, life will become, in the phrase used by English philosopher Thomas Hobbes (1588-1679), even more "nasty, brutish and short" than it already is at times.

If you think seriously about the proposals above, it should be clear that, in order to fight "climate change," the various national governments are going to have to interfere with (and degrade) our lives to a degree otherwise found only in wartime, and total war at that. Should democratic governments have this power?

Based on very dodgy "post-normal" scientific advice, which in turn is based almost entirely on flawed computer models, our governments are committing average citizens to a future of increasing poverty and a rather extensive loss of liberty. *This is the first time in history that democratic governments, outside of wartime, have deliberately tried to make their citizens poorer or less free.*

Evidence of government overreach is everywhere

The evidence of this governmental "green" overreach on climate is everywhere, and it impacts our lives in both big and small ways, always in the negative as far as individual freedom and prosperity is concerned.

For example, as of July 1, 2023, Canadians are currently paying 14 cents per litre for gasoline as a "carbon tax" that, we are told, is "revenue-neutral"—every cent of the tax will be returned to us as rebates. Sure it will.

But the real issue is that we shouldn't be paying this tax in the first place. What government in its right mind makes *energy*, which underlies *everything*, more expensive than it has to be, in the service of some vague plan to "fight climate change"? Higher energy costs equal higher living costs for everything we buy. A government in its right mind, one that is trying to ensure the prosperity of its people, tries to make energy as *inexpensive* as possible.

Sure, the gasoline carbon tax is currently "only" 14 cents a litre (for now). But the tax doesn't just affect car drivers filling up at the gas station. For example, truck drivers are also paying more, which means the cost of trucking goods is higher—is the Canadian government also going to give us a full rebate for the extra grocery costs caused by higher transportation costs getting food to supermarkets?

Then there are the many other "carbon taxes" paid by, for example, companies and industry in general to fight "climate change," costs that ripple through the economy and must eventually and inevitably be paid by the consumer. The price at the gas pump is just the most visible tip of the "carbon-tax" cost iceberg ... and a major fueler of inflation to boot.

But the government overreach doesn't stop with higher gasoline prices. Canadians have been told that after 2035 we will only be able to buy electric cars, whether we want them or not. EVs are a status symbol at the moment (there are Teslas everywhere in Greater Victoria, my city), but they are heavily subsidized to the tune of $2,500-$5,000 per vehicle by the federal government.[8]

To make matters worse, in the battle against "global warming," Ottawa is planning measures to decrease cattle farming and reduce use of nitrogen fertilizers,[9] as was done in Sri Lanka, a policy that collapsed that country's economy *in one year*.[10]

Dutch farmers are also under siege and fighting back against these types of regulations.[11] In Ireland, the government plans to order the sacrifice of 200,000 head of cattle in the service of the "green" gods.[12]

At a time when the world population is approaching eight billion, you'd think we'd aim to produce *more* food. Instead, the Canadian government and other governments are targeting farmers so they will produce *less* food. You can't make this stuff up.

And not just farmers—Ottawa is also coming for the oilfields workers and many others involved in the fossil-fuel industry and its network affiliates (and ultimately we are all part of this network), with promises of a "just transition" to "green" jobs. The government rhetoric strongly implies these "green" jobs will be just as good as the fossil-fuel ones; does anybody really believe this?

At the local level, here in the City of Victoria, contractors have been told they will no longer be able to install natural-gas heating in new buildings after 2025.[13] Shouldn't that be the builders' and home-buyers' choice?

Similarly, plastic straws, single-use plastic utensils and plastic bags have been banned by Ottawa and local councils, as if this policy will actually have any measurable impact on "global warming."

The examples of government interference in our daily lives in the service of "fighting climate change" go on and on, well beyond what should be permissible in a democratic system, and most of this meddling will have virtually zero impact on climate.

Given that we're not currently in "wartime," except in the fertile minds of the climate alarmists, one has to ask: are these sacrifices of personal liberty to fight an invisible foe warranted? I'd say not.

This is, in effect, what is being asked of us: to "fight" a "war" against an adversary ("climate change") that exists primarily in climate models. Yet this so-called "crisis" is so amorphous that a UN organization (the IPCC) needs to produce reports every five years to persuade us of its existence.

These reports are then magnified by environmentalists and political ideologues to make it appear we are facing a "wartime" emergency similar to the aggression of Hitler's Germany and Imperial Japan in the 1930s and '40s.

What we are asked to give up

Here's what a Canadian environmentalist suggests as the sacrifices he would like us all to make for the greater good, while wondering (rightly) if we are prepared to make these sacrifices:

> Parents and grandparents say they love their children (and grandchildren) and would do anything for them. OK, if that's true, does "anything" include giving up the Cadillac SUV for a hybrid VW, forgoing the annual migration to Hawaii or Barbados for a camping trip in the woods, and abandoning plans to double the size of the house from 3,000 to 6,000 square feet?
>
> Granted, most Canadians can't afford such luxuries, so let's consider some more modest measures as well. How about eating less or no meat, taking public transit instead of driving, shopping less, switching off lights and the computer, and turning down the AC? What do you think?[14]

I think the average person, if *candidly* offered this choice of a diminished lifestyle by the government, would say "No, thanks. I'll take my chances with 'global warming'!"

Personally, I'm not willing to risk the material well-being of my children and grandchildren based on IPCC reports that employ "post-normal" science and are hyped by the environmental radicals who produce the scary Summaries for Policymakers and media reports.

But the situation gets worse. The alarmists not only want to overturn almost every social, political and economic institution of liberal Western civilization—they also want to rewire human nature to make it happen, as we'll see in Chapter 16.

Notes to Chapter 15

---

1 "Absolute Zero." *UK Fires*. Available online at https://ukfires.org/absolute-zero.

2 George Monbiot: *Heat: How to Stop the Planet from Burning*. Toronto: Anchor Canada, 2007, p. xv.

3 United Nations Environmental Program, "Making Peace with Nature: A scientific blueprint to tackle the climate, biodiversity and pollution emergencies." 2021, p. 15. Available online.

4 "Making Peace," p. 29.

5 "Making Peace," p. 29.

6 "Making Peace," p. 29.

7 Michael Shellenberger & Ted Nordhaus, *Break Through: From the Death of Environmentalism to the Politics of Possibility*. Houghton Mifflin Harcourt, 2007. For a summary of *Break Through* and other authors with the same optimistic environmental message, see also my book, *False Alarm: Global Warming Facts Versus Fears*, Chapter 9, "A Brief History of Progress, and Why We Don't Believe in It." I also recommend Bjorn Lomborg's *The Skeptical Environmentalist* and Gregg Easterbrook's *A Moment on the Earth*. Hans Rosling's *Factfulness* also provides a welcome alternative to the doom-and-gloom view that claims our civilization is collapsing. The Rosling website, gapminder.org, should be an eye-opener for those who believe the world is getting worse rather than better.

8 Ottawa is offering rebates of $2,500-$5,000 toward the purchase of EV cars, and up to $200,000 for heavy vehicles like trucks. See "Zero emission vehicles," Government of Canada website.

9 Jim Weismeyer, "New policy forces Canadian producers to cut back on fertilizer." *AGWeb Farm Journal,* August 22, 2022. Available online.

10 See Wikipedia, "Sri Lankan economic crisis (2019-present)," and especially the section on "Agricultural crisis."

[11] Wikipedia, "Dutch farmers protests."

[12] Frank Bergman, "Ireland to slaughter 200,000 healthy cows to 'fight global warming'." *Slay,* June 3, 2023.

[13] *Times Colonist*, "No natural gas for new Victoria buildings by 2025." Aug. 9, 2022. Available online.

[14] Nicholas Reid, "Canadians are not willing to sacrifice for climate." *Vancouver Sun*, April 21, 2023.

Chapter 16: Less freedom—and loving it!

Oh, Kitty, how nice it would be if we could only get through into Looking-Glass House! I'm sure it's got, oh!, such beautiful things in it! Let's pretend there's a way of getting through into it, somehow, Kitty. Let's pretend the glass has got all soft like gauze, so that we can get through. Why it's turning into a sort of mist now, I declare! It'll be easy enough to get through.

As noted in Chapter 15, Net Zero by 2050 will mean giving up many or most of the rights and freedoms we associate with a liberal democratic society and Western civilization in general, including *psychological* freedom—the freedom to think differently than Big Brother.

In terms of rights and freedoms, a 2023 report from the Public Health Agency of Canada—in other words, a report with the backing of the federal government—offers a consistently alarmist analysis of the climate situation as it pertains to health, based on interviews with about 30 health professionals.

One of these professionals is quoted (anonymously), with the implicit approval of the report's three authors (and therefore of the federal government), as follows:

> Ultimately, there are three core values in western society, and for that matter, in global society, that have to change. One core value is about growth and materialism. *The second core value is liberty and individualism, which has to be rethought because the kind of individualism that is preached by neo-liberals is part of the problem.*
>
> It advances the individual over the collective, it says as long as I get what I want, bugger you, and it leads to a huge

> number of problems, and *it undermines the collective process.* A third core value that has to change is around our separation from nature, and [that] somehow, we're separate and apart from nature.[1] [emphasis added]

Bad enough that we are expected to give up "growth and materialism"; we also need to give up "liberty and individualism," two "core values" that have made liberal Western societies the richest and freeest in the world, a wealth and freedom, incidentally, that is gradually spreading to other parts of the globe (see Chapter 10). "Liberty and individualism" *are* Western civilization; without them there is no Western civilization.

Whenever reason and individual freedom are lost, humanity suffers. I fear we are going down the same path when climate alarmism declares war on the values that have made Western lives the best in history.

What Imperial Japan and Nazi Germany couldn't do from without during World War II—destroy Western democratic civilization—Net Zero attempts from within, and it's a truism that nations are almost always destroyed from within, rather than from without. We should be very concerned at how close this latest attempt to destroy Western civilization from within is to succeeding.

Personally, I prefer the much more rational view of Western civilization by historian Victor Davis Hanson, who has written:

> Western civilization has given mankind *the only economic system that works*, a rationalist tradition that alone allows us material and technological progress, the sole political structure that ensures the freedom of the individual, a system of ethics and a religion that brings out the best in humankind.[2] [emphasis added]

Similarly, Robert Heilbroner was a Marxist economist who (somewhat grudgingly) realized that *Marxism did not work and Western capitalism did.* The result was a 1989 essay in *The New Yorker* magazine entitled "The Triumph of Capitalism." It began:

> Less than seventy-five years after it officially began, the contest between capitalism and socialism is over: capitalism has won. The Soviet Union, China, and Eastern Europe have given us the clearest possible proof that *capitalism organizes the material affairs of humankind more satisfactorily than socialism:* that *however inequitably or irresponsibly the marketplace may distribute goods, it does so better than the queues of a planned economy*; however mindless the culture of commercialism, it is more attractive than state moralism; and however deceptive the ideology of a business civilization, it is more believable than that of a socialist one.[3]

If we're willing to give all this up—material prosperity, rationality, a political structure based on freedom of the individual, and so on— to fight a nebulous "climate crisis," so be it. But we shouldn't throw away our democratic system and individual freedoms with our eyes closed, based on the illusion that we can somehow "stop climate change," as the various governments want us to believe.

'Reenchanting' the world

On my shelf is a 1981 book by historian Morris Berman entitled *The Reenchantment of the World* that nicely encapsulates this climate-alarmist world view, although it appeared before climate alarmism had taken flight.

Berman is highly critical of Western logical, rational, empirical, scientific thinking because this thinking divides up the world, so that our experience loses (or appears to lose) its primordial "unity" with nature/reality. The argument echoes Al Gore's lament that

> I believe our civilization is, in effect, addicted to the consumption of the Earth itself. This addictive relationship distracts us from the pain of what we have lost: *a direct experience of our connection to the vividness, vibrancy, and aliveness of the rest of the natural world.*[4] [emphasis added]

As part of the European rationalist Enlightenment, Berman notes, "we hammered out a new way of perceiving reality," shifting from "quality to quantity, from 'why' to 'how'."[5] In other words, we have lost the sense of the "enchantment" of the world that less-rational, less-scientific peoples have (ostensibly) had, to (for Berman) our profound loss. Berman's hope is that we can get this "enchantment" back.

If we do, he promises a much better life for all of us! He writes:

> Such a society would be preoccupied with fitting into nature rather than attempting to master it. ... Technology will no longer pervade our consciousness and its presence will be more in the form of crafts and tools, things that lie within our control rather than the reverse.
>
> We will no longer depend on the technological fix, whether in medicine, agriculture, or anything else, but instead favor solutions that are long-term and address themselves to causes rather than symptoms.[6]

This future society would be highly decentralized (goodbye, cities!):

> Characteristic of such decentralization are community hospitals and food co-operatives, the cultivation of neighborhood spirit and autonomy, and the elimination of such destroyers of community as television, automobiles, and expressways.
>
> Mass production will yield to craftsmanship, agri-business to small, organic, labor-intensive farming, and centralized energy sources—especially nuclear power plants—to renewable energy options appropriate to their own regions.... One will not have a career but a life....[7]

In short, we would be returning to a version of the Middle Ages, which was clearly a much better time than today because, amidst the squalor, poverty, superstition, religious persecution, disease, and early death, at least we were "enchanted"! It's "enchanted" people, after all, who believe in, say, witches.

When I first read Berman's book, and especially the paragraphs above, I thought, "How can any intelligent academic historian, much less any intelligent person, believe this nonsense, much less write a book about it?" The answer is simple: because Berman is under the spell of Marxism, either in name or spirit.

Now, more than 30 years after the collapse of the Soviet Union in 1991, after we've seen how well (being ironic here) communism continues to operate in countries like China and North Korea, it is astonishing that Marxism still has any credibility in the modern world. And yet, idealism will always exist, and idealism often attaches itself to the vision of Karl Marx.

This is why Berman's wonderful new "enchanted" world—which is also the world envisioned by the United Nations ("Making Peace with Nature") and the Canadian government ("What We Heard: Perspectives on Climate Change and Public Health in Canada"), among others—sounds pretty much like (or, really, identical to) Marx's imagined ideal community once communism had been fully achieved:

> In communist society, where nobody has one exclusive sphere of activity but each can become accomplished in any branch he wants, society regulates the general production and thus makes it possible for me to do one thing today and another tomorrow, to hunt in the morning, fish in the afternoon, rear cattle in the evening, criticize after dinner, just as I have a mind, without ever becoming hunter, fisherman, shepherd or critic.[8]

In Marx's vision as well, one will "not have a career but a life." Of course, some unenlightened people actually enjoy their careers while still having "a life"....

'Enslaved' by machines—the 'tragedy of the scientific era'

A more recent version of this myth of an un-"alienated" (to use Marx's term for the fully "conscious" class warrior) and therefore "enchanted" individual and society comes from Canadian writer Andrew Nikiforuk in his 2012 book *The Energy of Slaves: Oil and the New Servitude*.

As with Berman's book, I was astonished that any rational human being could write this idealistic nonsense, much less find a publisher willing to print it.

Nikiforuk compares the moral evil of human slavery to the, for him, equally moral evil of our "enslavement" by machines, industry, and technology.

That is, just as the Southern plantation owners were morally corrupted by having their work accomplished by human slaves, so we today are morally corrupted by our dependence on machines to do much of our work. "Once dependent on the energy of slaves, we are now slaves to petroleum and its masters," Nikiforuk writes.[9]

He calls our dependence on technology the "tragedy of the scientific era."[10] Indeed, who could possibly support a "scientific era" that has doubled human lifespans, overcome most of the diseases that have plagued humankind, given us modern sanitation, and overall made most of us materially better off and much less dependent on the whims of nature than those in the "pre-scientific" era. Truly "tragic"!

On the plus side, from a more rational point of view, Nikiforuk recognizes that "renewable" energies alone won't keep the lights on and the trains running: "All currently proposed wind, solar, thermal, and biomass projects still add up to less energy than the current global consumption of 1.5 terawatts."[11]

Nikiforuk expands on this observation in an article that explains why he has become a "renewable-energy skeptic" regarding our proposed Net Zero, all-electric future.

In the article, Nikiforuk shows rather conclusively that the planet doesn't have all the minerals, particularly "rare earths," needed for a full Net Zero electrification of energy use (e.g., everyone driving electric cars), as the alarmists want.

Worse, for Nikiforuk, the attempt to find these minerals will require *even more mining than now* and therefore will *increase* what he sees as the human despoilation of the planet. Nikiforuk writes:

> The average smart phone contains at least 40 elements from the periodic table including cobalt and six rare-earth minerals that make the screen glow. *The average electric car uses six times more critical minerals than a combustion car.*

> An onshore wind plant needs *nine times more mineral resources than an equivalent gas-fired power plant.* An e-bike is more mineral-intensive than an ordinary bike. And so on. Renewables haven't just accelerated the demand for rare-earth minerals but a variety of base metals such as copper, silver and cobalt.[12] [emphasis added]

He goes on:

> Every electric vehicle contains about *75 kilograms of copper or three times more than a conventional vehicle.* A single wind turbine generally contains 500 kilograms of nickel. That nickel requires 100 tonnes of steelmaking coal to be refined. And every crystalline silicon solar panel contains 20 grams of silver paste. It takes 80 metric tons of silver to generate approximately a gigawatt of solar power. (In power terms that's equivalent to 9,000 Nissan Leafs.)[13] [emphasis added]

So Nikiforuk is on the horns of a dilemma. He hates modern fossil-fuel-based capitalism with its individualism, dependence on technology, and aggressive desire for growth. But he also hates the Net Zero future since it, too, will need mining and other "tragic" scientific evils, just like capitalism but with a "green" veneer!

His solution? It is not to find new energy sources actually capable of sustaining our current lifestyle (nuclear, anyone?). Instead, Nikiforuk wants to *reduce* our lifestyle choices so we don't need as much energy, which means going back, like Berman, to an "enchanted" society much like that of the Middle Ages.

I'm not exaggerating here. Nikiforuk wishes us to aim for the modest and self-sufficient lifestyle of ... wait for it ... the Benedictine monks of the Middle Ages. Benedictine monasteries were "autonomous and they generally operated on a small scale."[14] They "demonstrated that scarcity did not mean poverty; that abundance did not mean affluence; that progress did not mean momentum; that goodness did not mean growth."[15]

With breathtaking optimism, Nikiforuk offers his vision of the future (which he shares with Berman and Marx in almost identical terminology), once we've seen the error of our oil-using ways:

> Around the world, families and groups of individuals are walking away in ever-growing numbers from petroleum and the inanimate slave culture of frantic consumption. They are exchanging quantity for quality and relearning the practical arts. Those seeking liberty eat slowly, travel locally, plant gardens, work ethically, build communities, share tools, and eschew bigness in economic and political life.[16]

Post-'green' life: Nasty, brutish and short

Sounds great! Unfortunately, the Bermans and Nikiforuks, UNEP, the Canadian government and all other "green" governments, have no coherent, rational, do-able plan for keeping today's eight billion people alive (much less keeping the lights on) once we've stripped ourselves of the (relatively) cheap energy of fossil fuels in the name of fighting "climate change."

Millions, perhaps billions, of us will die and/or suffer because (as even Nikiforuk accepts, having become a "renewable-energy skeptic") there is no "renewables" plan, no Net Zero plan, that can possibly succeed in keeping all eight billion of us alive and functioning as a society except in the imaginations of the alarmist wordsmiths (and, again, all this sacrifice for a tiny gain in "global cooling").

We will also be saying goodbye to "progress," another hallmark of Western civilization (it is richly ironic that the anti-growth, environmental neo-Marxists call themselves "progressives"). Western civilization has developed technology sophisticated enough to put humans on the Moon, but for the alarmists we apparently aren't smart enough to adapt to 2-4 degrees Celsius of warming (assuming that even happens).

Under the Net Zero "enchanted" plan, humanity would never again go to the Moon, for example, and we will never visit Mars. As a species we shall cease from exploration and live in small, static communities, more or less as our ancestors did in the Middle Ages or earlier (let's hope we don't go back to the Stone Age!). This is the future planned for us by climate alarmists.

Meanwhile, while we're huddling in our small, "sustainable" communities, Communist China, which will not subscribe to Net Zero until its economy is fully "developed" (i.e., whenever China feels like it), is building one or two coal-fired power plants a week.[17] And China will continue along the path of building its economy using fossil fuels until it is *the* global superpower. At that point it will dominate our "Western," small-scale, "enchanted" communities at will, without a shot being fired.

This is where Net Zero is leading liberal Western civilization, for those who care to look ahead—a world dominated by a totalitarian state. Orwell's prediction will have come true: "If you want a picture of the future, imagine a boot stamping on a human face—for ever."[18]

A gradual transition is needed

Thinking realistically, rather than in Net Zero, Alice in Wonderland terms, if we are to transition to a post-fossil-fuel society, we will need to reduce our use of fossil-fuels *gradually*, over many decades, not two or three (Net Zero by 2100 or later, but certainly not 2050). We can't just go "cold turkey" on fossil fuels and hope for the best, which is the Net Zero by 2050 plan.

But this "cold turkey," "hope for the best" approach is what is being promoted for Canada and the rest of the world by the climate alarmists, while beguiling us with visions of the "greener," more "enchanted," less-"alienated" future we will all enjoy so much more than what we have now (like those lucky Benedictines, for example!).

What about people in the developing countries, tens of millions of whom *want* the prosperous lifestyle of the West—want it enough to cross borders as immigrants and refugees, sometimes at the risk of their lives?

Well, in alarmist theory these unfortunates will get a bigger piece of the diminishing material pie. In reality, tough for them. Any system that aims to keep the rich from getting richer will also keep the poor from getting richer. The only beneficiaries of Net Zero by 2050 will be the insulated elite who are proposing all this non-prosperity in the first place (World Economic Forum, we're looking at you!).

Meanwhile, the ultimate agenda of the climate alarmists is pretty clear. When former Canadian environment minister Christine Stewart observed, "No matter if the science is all phony, there are collateral environmental benefits," she revealed that the "collateral benefits" she and her alarmist colleagues are so enthusiastic about is something like this neo-Middle Ages lifestyle. We'll have less "stuff," we'll be less free, less individualistic, but by God we'll be happier!

Maybe. Or maybe not. This type of Utopian vision has never worked before in human history. But this time it will, the alarmists tell us, if only we're willing to give up our dependence on fossil fuels and other forms of energy and embrace doing less with less.

Just as importantly for the alarmist program, we need to change our human nature to make this transformation work.

Changing human nature: 'Inner transitions'

Inner "transformational" change of human nature is, incredibly, an element of the IPCC's latest series of reports (2021, AR6). The AR6 Working Group III report notes:

> While *transitions* involve "processes that shift development pathways and reorient energy, transport, urban and other subsystems," ... *transformation* is the resulting "fundamental reorganization of large-scale socio-economic systems." Such a fundamental reorganisation often requires dynamic multi-stage transition processes.[19] [emphasis the authors']

Translating this post-modernist gobbledygook into plain English, climate alarmists are going to interfere, big time, in our society's "socio-economic systems" (goodbye, free markets!).

To make this "fundamental reorganization" work, human nature will also have to change, but the alarmists have that covered. We'll undergo a psychological "inner transition" so we won't miss what we're losing. In fact, we'll be much happier with much less! The IPCC report goes on:

> The starting point for this virtuous circle is *inner transitions*. Inner transitions occur within individuals, organizations and even larger jurisdictions that alter beliefs and actions involving climate change. *An inner transition within an individual typically involves a person gaining a deepening sense of peace and a willingness to help others, as well as protecting the climate and the planet.*
>
> Inner transition can imply that individuals become sympathetic to concerns that include climate issues and values connected to nature. For instance, they may include a desire to become a steward of nature; "live according to the principles of integrated sustainability"; "achieve the good life"; or protect the well-being of other living creatures.[20] [emphasis added]

Another word for creating "inner transitions," just for the record, is "brainwashing." Brainwashing is an attempt to change not just the way we *behave* but the way we *think*.

As history has taught us, for totalitarians it's not enough for the people to *behave* as if they love Big Brother; they have to *actually* love Big Brother. Brainwashing worked on Winston Smith in Orwell's *1984* and maybe, in the brave new world the alarmists are designing for us, we can, like Smith, also learn to love Green Big Brother with the innermost fibre of our being!

As a point of interest, part of the IPCC's mission statement is that the organization is "policy-neutral, never policy-prescriptive."[21] This "inner transition" the IPCC is proposing sounds pretty darned "policy-prescriptive" to me, so once again the IPCC says one thing and does another.

Summing up: What the IPCC and its alarmist allies push for is not just a radical change in the Western-style social and economic system. They are aiming for nothing less than a rewiring of human nature along Marxist lines so we will not only accept lower living standards and reduced personal, political and economic freedom, but also *love* it!

Attempts to change human nature didn't produce a perfect society in Soviet Russia, haven't worked in Communist China or North Korea. Why would it work for the Western democracies?

If we don't want this totalitarian outcome then we, as citizens, need to look more carefully at the "green," "inner-transitional" lifestyle proposed for us in the name of "fighting climate change." It's a lifestyle that most of us would instantly reject if this utopian future was explained clearly and honestly by our governments (which it isn't), in terms of both financial cost and the "wrenching changes" (material *and* psychological) involved.

In the next chapter, I'll present what I hope is a *rational* approach to whatever the planet wants to throw at us in terms of "climate change" (prediction: the future climate will be nowhere near as bad as the doomsayers wish us to believe).

Notes to Chapter 16

---

1 Heather Castleden, et al., "What We Heard: Perspectives on Climate Change and Public Health in Canada." *Public Health Agency of Canada,* 2023. Available online.

2 Victor Davis Hanson, *Carnage and Culture: Landmark Battles in the Rise of Western Culture.* New York: Anchor, 2001, p. 455.

3 Robert Heilbroner, "The Triumph of Capitalism." *New Yorker*, Jan. 23, 1989, p. 98.

4 Al Gore, *Earth in the Balance: Ecology and the Human Spirit.* Toronto: Penguin Books, 1993 (1992), pp. 220-224.

5 Morris Berman, *The Reenchantment of the World.* Cornell University Press, 1981, p. 45.

6 Berman, *Reenchantment*, p. 276.

7 Berman, p. 276.

8 Karl Marx, *The German Ideology*, "Feuerbach," "1. History." Moscow: Progress Publishers, 1968, p. 45.

9 Andrew Nikiforuk, *The Energy of Slaves: Oil and the New Servitude.* David Suzuki Foundation/Greystone Books, 2012, p. xii.

[10] Nikiforuk, *Energy of Slaves,* p. 141.

[11] Nikiforuk, *Energy,* p. 225.

[12] Andrew Nikiforuk, "The rising chorus of renewable energy skeptics: The green techno-dream is so vastly destructive, they say, 'we have to come up with a different plan'." *The Tyee,* April 7, 2023. Available online.

[13] Nikiforuk, "Rising chorus"

[14] Nikiforuk, *Energy of Slaves,* p. 246.

[15] Nikiforuk, *Energy,* p. 248.

[16] Nikiforuk, *Energy,* p. 249.

[17] Robert Lyman, "China is winning the climate policy game." *National Post,* March 28, 2023.

[18] George Orwell, *1984*. Penguin Books, 2008 (1949), p. 280.

[19] IPCC, AR6, Working Group III, Chapter 17, Section 17.2, p. 1734. Available online. Cited in Roger Pielke, Jr., "The Political Agenda of the IPCC." *Substack,* May 15, 2023. Available online.

[20] IPCC, AR6, WGIII, Section 17.2.3, p. 1737.

[21] IPCC, "What is the IPCC?" Available at https://www.ipcc.ch/lenguages-2/English.

Chapter 17: Piecemeal engineering vs. utopian social engineering

Well, some go this way, and some go that way. But as for me, myself, personally, I prefer the short-cut.

Because "global warming" is a huge, amorphous problem, it has many "faces." For example, at the time of writing (summer/fall 2023), a local newspaper weekly column was entitled "Climate action needs a greater sense of urgency" because, the columnist opined, 2023 is having the "worst wildfire season in the past 20 years."[1]

For this columnist the upsurge in fires was due, of course, to "climate change," hence the need for urgency (although, wasn't the situation just as "urgent" 20 years ago, when fires were just as bad or worse? And yet, somehow, here we still are).

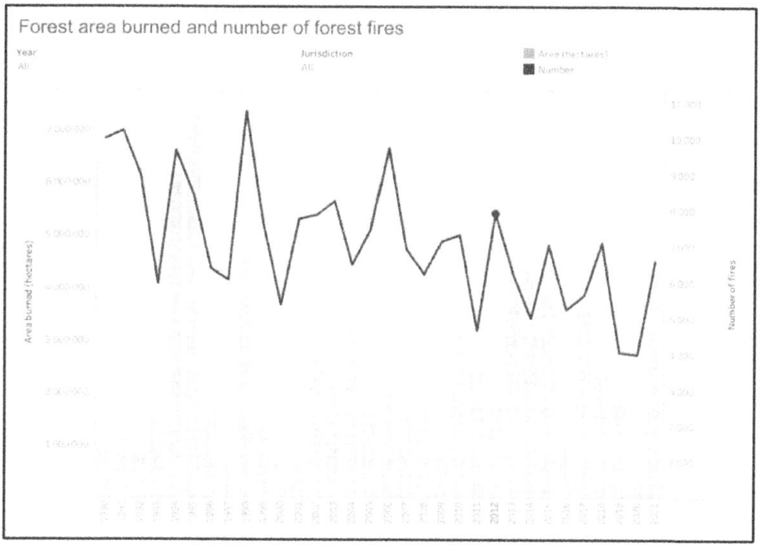

Figure 17.1: Number of forest fires decreasing over 30 years. Source: National Forestry Database

Oddly, though, the number of fires in Canada has actually gone *down* over the past 30 years, according to the National Forestry Database (see Figure 17.1, above).

If forest fires are one of the "faces" of global warming, shouldn't the number of fires have gone steadily *up* for those 30 years of previous "warming," rather than down? Or is it possible, as this book has suggested, that there hasn't been much warming at all in these 30 years and that therefore the warming that has occurred cannot be due to increased CO2?

Another "face" of climate change in British Columbia and globally in the summer of 2023, as we saw in Chapter 5, is drought caused by "record" temperatures, yet another (alleged) product of "anthropogenic global warming."

The very clear implication for the public is that the planet has reached untold ("record") warming heights that will soon burn us into crisps. But today's warming is only a "record" since 1850 or so, when we developed instruments to accurately measure temperature. Is it possible the planet was warmer before 1850?

If the planet *was* warmer before industrialization (so today's temperatures aren't the "record" after all), what caused this higher previous warmth—it couldn't have been fossil fuels. And how could our less-technological ancestors possibly have survived this "extreme" heat?

A little research reveals some interesting facts about, for example, the current "record-breaking" drought in British Columbia.

In 2016, University of Victoria researchers reported, based on tree-ring data, that in the past 350 years the province had experienced at least 16 droughts that were *worse* than the drought at that time. Oddly, many of these earlier droughts occurred centuries before fossil-fuel emissions were an issue.[2]

Logically, these 16 earlier, worse droughts must have been the result of *warmer* temperatures than today. If the planet was warmer in years before the instrumental temperature record and somehow humanity survived, then today's temperatures are still within "natural variation"—there's no need to fear that the current warming will make us extinct.

Could the *natural* (non-fossil-fuel) causes of these earlier, warmer temperatures, which caused earlier, worse droughts, possibly still be active today? As noted in Chapter 5, climate alarmists point to 2023's "record" temperatures, forest fires, droughts, etc., as definitive proof that their side of the mirror reflects Reality[3] and the "deniers" should just get on board.

For alarmists, 2023's "record" heat couldn't be, for example, a result of the massive Tonga undersea eruption that increased the amount of water vapor in the atmosphere by 10-15 per cent (see Chapter 5).[4] Or about the large El Niño just getting up to speed in 2023.

Or any of a dozen *natural* causes that all work together to create "climate change," including the 16 more severe droughts over the past 350 years (and no doubt much earlier).

Nor is it possible, for alarmists, that reducing CO_2 levels might have *absolutely no impact* on these high temperatures, now or in the future, thanks to the carbon band saturation noted in Chapter 6.

'Climate change' has many 'faces'

In other words, "climate change" has many faces. For some, at some times, "climate change" means more (and more-intense) forest fires; for others more floods; for others more droughts; for others more cases of extreme heat; for others sea-level rise threatening cities; for the ultra-alarmists, all of the above, and so on.

As used by climate alarmists, "global warming" is a one-size-fits-all explanation for almost any weather phenomenon they choose, with the risk that, as Karl Popper put it, a theory that explains everything explains nothing.

None of these "faces," alone or in combination, are new; humans have been dealing with them for thousands of years, mostly with success. Why? Because we have found affordable, engineering-based solutions for all these so-called "climate-change"-related issues.

Our technological toolkit contains better fire-management practices; better flood-control measures (the 2021 Sumas Prairie flooding in

B.C. was caused by failure to maintain the embankment system, not "global warming"); better channeling and use of water resources to reduce drought; cooling measures (such as air-conditioned community centres) to protect vulnerable people for the few days a year of "extreme heat" (the week-long B.C. "heat dome" of June 2021 comes to mind).

Thinking the unthinkable: Moving cities

As for sea-level rise threatening cities—we can protect them with dikes and embankments, as the Dutch have done with great success. Or, thinking the unthinkable, as used to be said of nuclear war, in a worst-case scenario we could move and rebuild threatened cities away from the flood zones.

Again thinking the unthinkable: what if desert or near-desert areas like, say, Texas or Arizona, become too hot to be habitable (for example, Neil Stephenson's 2021 science-fiction novel *Terminal Shock* suggests that people in Texas would have to wear air-conditioned "earth suits" to survive severe global warming there). Do we dismantle Western civilization, as the alarmists wish, or do we move those threatened cities and their people to more climatically moderate areas?

If this sounds impossible, consider that during the Second World War hundreds of major European and Asian cities were virtually obliterated—and then rebuilt when the war was over.

For example, 60 per cent of Japanese cites were mostly wiped out; 40 per cent of Germany's built-up areas were destroyed.[5] These cities, and cities in many other war-ravaged nations, were rebuilt, sometimes almost from scratch—for example, almost 90 per cent of Warsaw, Poland, was destroyed, and much of it restored. Similarly, Stalin moved much of Soviet Russia's heavy industry across the Ural Mountains as the Nazi invasion threatened Moscow.

The difference between WWII and now is that we have time—decades or even centuries—to react to rising sea levels, if sea levels are actually a real problem (instead of just another alarmist bogeyman), or droughts, or any of the other predicted evils of "global warming."

In the meantime, perhaps we should ask the residents of, say, Chicago or Winnipeg how they'd feel about demolishing their economies with Net Zero measures to save New York or Halifax from sea-level flooding, or Houston from extreme drought. No? Thought not.

Incredible as it sounds, building new cities while preserving the global economy *is a much more sensible response* to extreme sea-level rise or extreme drought (if they are an existential problem, and that hasn't been the case so far) than destroying the global economy through a Net Zero policy that won't even work if CO2 is not the "control knob" of climate.

Critics might suggest that evacuating, say, New York City or Houston, Texas, would mean abandoning many billions of dollars of infrastructure and displacing millions of people. But, as we've seen, *this is exactly what the climate alarmists want to do even if global warming isn't a threat.* Recall from Chapter 15 the UNEP report that notes:

> Existing infrastructure and built capital provide system inertia that can make change difficult and *incur short-term costs, especially if change involves the premature retirement of capital stock.* ... Individuals and organizations can also *oppose change that disrupts their livelihoods, market share and revenues, or that otherwise appear unfair.*[6] [emphasis added]

For the Houston or New York refugees there will be massive "short-term costs and premature retirement of capital stock"—but *no more than the alarmists are planning to impose on Texans and New Yorkers (and everyone else) anyway* with their Net Zero plan.

Why not, then, at least keep Western civilization functioning in the areas of the world that are still habitable. (I repeat, this global-warming scenario is extremely unlikely, but we're thinking the unthinkable here.)

Piecemeal versus utopian social engineering

Philosopher of science Karl Popper called practical solutions to problems like global warming "piecemeal engineering," which means tackling problems as they come up, as opposed to what he called "utopian social engineering," which aims at some broader, overarching social goal.

For Popper, the sensible reformer may have an ideal blueprint of the society he/she wishes to create,

> but he will be aware that perfection, if at all attainable, is far distant. The piecemeal engineer will, accordingly, adopt the method of searching for, and fighting against, *the greatest and most urgent evils of society, rather than searching for, and fighting for, its greatest ultimate good.*[7] [emphasis added]

Elsewhere he makes the point another way:

> *Work for the elimination of of concrete evils rather than the realization of abstract goods.* Do not aim at establishing happiness by political means. *Rather aim at the elimination of concrete miseries.*"[8] [emphasis added]

This means that realism rather than idealism should be the politician's watchword when it comes to climate (or any issue). The purpose of government isn't to make us "happy"—that's our *personal* goal. The purpose of government, for Popper, is to tackle "concrete miseries."

If you have raccoons in the attic, the "solution" isn't burning the house down to get them out, as the climate alarmists wish to do; the solution is calling the exterminator. In climate terms, this means tackling the problems that "global warming" might bring up through a "piecemeal-engineering" approach.

'Incremental trade-offs' vs. 'categorial solutions'

Economist Thomas Sowell describes the same approach a slightly different way in his 1980 book *Knowledge and Decisions.*

He points out that "no given individual or manageably sized group is likely to have sufficient knowledge to be consistently right the first time in their decisions." Therefore, we need "feedback mechanisms" that, in the short term, inform us of whether our actions are succeeding or failing in achieving our goals.[9]

To get these shorter-term feedbacks, Sowell says, we need to rely on "incremental trade-offs" [i.e., piecemeal engineering] rather than attempting "categorial solutions"[10] [i.e., utopian social engineering].

For example, in the current Net Zero, "categorical solution" approach to "global warming," how do we know if we are succeeding or failing? We don't, at least so far.

In fact, given that none of the world's nations are meeting their "targets" for curbing carbon emissions, and that some huge economies like China and India aren't even trying, it's more likely we are failing in our efforts to reduce warming. But how can we be sure? Especially since the climate stubbornly refuses to warm as much as the models say it should?

Is this "temporary slowdown" in warming (the "hiatus") caused by humanity's anti-carbon efforts, or by variations in nature? Based on their reaction (noted in Chapter 7), even alarmists believe natural variation is at fault for the 2000-2014 "hiatus."

Often in the utopian "categorical" approach, as Sowell notes, "the point is reached *where trivial amounts of one thing are gained at the cost of devastating losses of another*."[11] [emphasis added] As we saw in Chapter 14, Net Zero could mean a "devastating" 10 per cent loss of GDP (or more) to reduce the temperature a "trivial" fraction of a degree Celsius.

On the other hand, if we adopt the piecemeal engineering, "incremental trade-off" approach, we can get a handle on, say, increased forest fires, which may be due to "global warming" or maybe not. In the end, the *cause* of the fires doesn't matter if we see, through incremental feedbacks, that our efforts have reduced forest-fire damage (or not). And the same for the other "faces" of "global warming."

In the same vein, Sowell makes another important point: how we approach these incremental trade-offs is crucial.

Do we tackle problems through a methodology of "systematic authentication" that "involves a testing of the logical structure of a theory for internal consistency and a testing of the theory's results for external consistency with the observable facts of the real world"?

In other words, do we use the methodology of *science* (and Sowell means traditional science), which compares theory against reality through testing? Or, he writes, do we approach problems through a methodology of "consensual approval," meaning

> the approval of the general public as of a given time, or the approval of some special reference group—a social class, a religious sect, an ideological movement, etc. ... Sometimes the elitism implicit in such a [consensual] position can be tempered by depicting the idea in question ... as beneficial to a broad sweep of mankind outside the group, so that the group is only a temporary surrogate for a larger constituency which will ultimately approve the idea."[12]

The "consensual" approach—and keep in mind that *Knowledge and Decisions* was written in 1980, almost a decade before institutionalized climate alarmism began with the founding of the IPCC—describes very exactly the methodology of the "Anointed," elite in-group that promotes climate alarmism through "post-normal" "consensus"-based "science," for our own good, of course.

Sowell also warns about the dangers of mixing the two methodologies:

> A problem does arise, however, when one method masquerades as the other—for example, when the results of essentially consensual processes choose to present themselves as scientific.[13] [emphasis added]

Sowell is describing with great precision—and again, he was writing in 1980—the methodology of "consensus," "post-normal" alarmist climate "science," which presents itself to the public as "real" science when it is nothing of the sort.

Piecemeal solutions not enough for alarmists

Obviously, for alarmist social reformers, piecemeal solutions aren't good enough. Their "solution" to forest fires, floods, heat waves, or whatever the most prominent "face" of global warming happens to be at any given time, isn't to focus on measures against these evils using piecemeal engineering.

As we saw in Chapter 14, their "solution" is to dismantle our technological civilization and economy by drastically reducing the use of fossil fuels—the categorical, utopian-social-engineering approach—based on the CO_2 "control knob" theory. The raccoons in the attic may be gone—so, success!—but so is the house.

In the process, as we saw in Chapters 15 and 16, the alarmists plan a radical program of utopian social engineering. This program will include giving up "liberty and individualism," a decline in our standard of living in favor of a smaller-scale, more communistic society, and even attempts at changing human nature (an "inner transition" that will make us all much, much happier). All these measures are excused by the dire necessity (alarmists believe) of confronting the supposedly overwhelming problem of "global warming."

But as any good psychologist will advise, you handle a large, amorphous problem (like, say, a patient's generalized anxiety) by breaking the problem down into manageable parts and tackling the parts.

Similarly, the "solution" to an apparently huge, nebulous problem like "global warming" isn't to dismantle Western civilization and replace it with a lower-tech socialist utopia that, like most utopias, will almost certainly make most of our lives worse.

The solution is to break the global-warming problem down to its component parts and then deal with each part through the "piecemeal engineering," "incremental trade-offs" approach that has worked well in the past, in part because we can measure how successful these trade-offs have been and change course if necessary.

Another benefit: the piecemeal approach allows us to continue to use fossil-fuel energy while we move *gradually and organically* to a new energy system, almost certainly nuclear, that can still sustain Western-style civilization.

Also, practically speaking, there's no guarantee that the Net Zero, utopian engineering approach will actually *fix* the problem—what if the current "global warming" *isn't* caused by rising CO_2 levels, as this book has argued (I hope persuasively)?

What if slashing CO2 levels, at enormous cost to the economy and our lifestyles, *doesn't* bring temperatures down, or at least enough to make a difference?

We'd have spent trillions of dollars, impoverished ourselves, and given up our individual freedom for nothing. As it stands, we have no guarantees at all that the Net Zero approach will actually have the "global-cooling" results it promotes.

By contrast, the piecemeal engineers just want to solve the immediate difficulty (flood, fire, drought, etc.). Another benefit, for Popper, is that the piecemeal-engineering approach will normally not cause a lot of social conflict, as the Net Zero approach does, since we can all agree that, say, fewer forest fires is a desirable goal.[14]

Let me give you an example from Victoria, B.C., of how the two visions conflict. In 2001 the region was experiencing one of its worst droughts in years (perhaps a "record" drought at that time), with the city's reservoir at its lowest level since 1900.[15] What to do?

One suggestion, a piecemeal-engineering approach, was raising the Sooke Reservoir dam, the main source of water for the region. However, local environmentalists, following the "Anointed," more moralistic green vision, argued against the dam project and preferred that residents practice more water conservation ("conserve" rather than "consume," as the current mantra goes).

Fortunately for the residents of Greater Victoria, the city decided to go with the more practical, piecemeal-engineering approach of raising the dam and the city has not been plagued with water shortages since.

The perfect is the enemy of the good. When looking for ways to cope with the problems that "climate change" appears to raise, we should always prefer the "good"—the pragmatic, piecemeal-engineering, incremental trade-offs approach—rather than the alarmists' utopian, "perfect" (but usually unrealizable) categorical solution.

Notes for Chapter 17

[1] Trevor Hancock, "Climate action needs a greater sense of urgency." Victoria *Times Colonist,* June 25, 2023.

[2] Wanyi Lee & Lisa Johnson, "B.C. has seen worse droughts than previously thought." *CBC News*, May 4, 2016. Available online. The article referenced is Bethany Coulthard, et. al., "Is worst-case scenario streamflow drought underestimated in British Columbia? A multi-century perspective for the south coast, derived from tree-rings." *Journal of Hydrology*, Vol. 534, March 2016, pp. 205-218.

[3] Bob Weber, "Something's changed: Summer 2023 screams climate change." *Canadian Press, Times Colonist,* August. 6, 2023.

[4] Ayesha Tandon, "Tonga volcano eruption raises 'imminent' risk of temporary 1.5°C breach." *Carbon Brief,* Jan. 12, 2023.

[5] For example, as historian Richard Overy notes in *Blood and Ruins,* his history of WWII, 60 per cent of Japan's cities were substantially demolished (p.370), and 40 per cent of Germany's build-up area (p. 586). And this doesn't count the many cities substantially destroyed outside these two countries, e.g., Manila and Warsaw.

[6] United Nations Environmental Program, "Making Peace with Nature: A scientific blueprint to tackle the climate, biodiversity and pollution emergencies," 2021, p. 29.

[7] Karl Popper, *The Open Society and Its Enemies, Vol. 1.* Princeton, NJ: Princeton Univ. Press, 1971 (1943), p. 148. Popper's approach has much in common with that of Thomas Sowell's dichotomy of the Anointed, unconstrained vision versus the constrained, more realistic vision.

[8] Popper, *Conjectures and Refutations: The Growth of Scientific Knowledge.* New York: Routledge & Kegan Paul, 2002 (first edition 1963), p. 485.

[9] Thomas Sowell, *Knowledge and Decisions.* New York: Basic Books, 1980 (1996 edition), p. xiv.

[10] *Knowledge and Decisions*, p. xii.

[11] *Knowledge and Decisions*, p. xii.

[12] *Knowledge and Decisions*, pp. 5-6.

[13] *Knowledge and Decisions*, pp. 5-6.

[14] Popper, *Conjectures and Refutations*, p. 483.

[15] Deborah Walker, "Victoria's strategy when facing the drought." *Environmental Science and Engineering,* March 2002.

Conclusion: Alarmist or realist?

Why, they're only a pack of cards, after all. I needn't be afraid of them!

In this book we've gone, a bit like Alice, though the looking glass, in this case the climate looking glass.

I've tried to show you, with lots of evidence, some of the illogical and sometimes absurd thinking, the belief in many "impossible things," on the alarmist side of the mirror, where "climate change" is considered a huge existential problem for humanity.

By contrast, I've also tried to show you, often with evidence you can check for yourself (hence a "do-it-yourself" guide), what's on the realist side of the mirror, where "climate change" is just another challenge to human ingenuity.

I hope I've given you many evidence-based reasons why I favor the realist side, even if that makes me a climate "denier" and therefore, in alarmist eyes, deluded, insane, a dishonest purveyor of "fake news," and/or "scientifically illiterate."[1]

Heck, a *Guardian* article divides "deniers" into four categories: the shill, the grifter, the egomaniac and the ideological fool.[2] Ouch! Hurtful!

And odd, because the climate "deniers"/realists I know personally are successful professionals in the fields of engineering, diplomacy, the military, politics, government, business, meteorology, and, yes, even science.[3]

They've got a wealth of practical experience and most have read widely and deeply on the issue of climate science. It's hard to believe they are deluded or dishonest or stupid or scientifically illiterate, as the alarmist propaganda machine claims.

On the 'realist' side of the mirror

Let's briefly review the argument in this book and see if the realist approach makes sense, or at least more sense than the alarmist side of Alice's mirror.

Temperatures 'flat-lined' outside of El Nino years

The planet is, overall, warming at the moment—or at any rate, not cooling. "Climate change" is "real," but for most of the past 70 years, since at least 1950, temperatures have been pretty much flat-line most of the time.

Any "warming" has been caused not by additions of human-generated carbon dioxide, at least not directly as the IPCC argues, but by several large El Niños that have been "ratcheting" the temperature upward every few years.

This "ratcheting" effect is very clear when we look at the actual temperature data since 1950 on Woodfortrees.org—since 1950 the temperature is pretty much flat between El Niños.

Almost no correlation between temperature and CO2 over time

Over deep time, meaning tens and hundreds of millions of years, the geological and historical temperature data shows almost no or weak correlation between temperature and carbon-dioxide levels.

For more recent times, including the past 50 million years, the past 650,000 years (Al Gore's wall chart), and the past few thousand years, before human industrial activity was a factor, the pattern is clear: when there *appears* to be a correlation, first the temperature changes, then the CO_2 level changes.

Granted, human activity has increased the planet's CO_2 levels in the past hundred years, but it's unlikely the laws of physics have changed to create a strong correlation between temperature and CO_2 just to accommodate alarmist ideology. There is every reason to believe climate alarmists are mistaken in believing that CO_2 levels today have a strong correlation with temperature.

Carbon band is 'saturated'

The failure of CO2 to warm the climate as much as the alarmist climate models predict (when the models are "hindcasted") is almost certainly due to the carbon band-saturation effect: the warming power of CO2 is strongest at low levels, up to say 200 parts per million, but falls off exponentially as CO2 increases.

If CO2 had the "control knob" warming power claimed by alarmist climate models, temperatures would be rising more *gradually and consistently* (not just during El Niños) than they actually are. In short, CO2 warming is "saturated."

Alarmists don't take this "saturation" into account in their predictions of catastrophic warming, on the assumption, again, that physics will modify itself to fit their beliefs.

By contrast: If the CO2 atmospheric band is "saturated," then CO2 levels would have to fall *drastically* to have any measurable effect on the climate—drastically enough to wreck industrial civilization, but also drastically enough take us back to a new Ice Age, when CO2 levels were around 200 ppm.

Planetary cooling would be far, far worse for human civilization than any possible warming we might face.

'Global warming' has many causes, most natural

Why is the planet warming at all? For the same mostly *natural* reasons that the planet has warmed in the past and will warm in the future, just as in turn the planet periodically *cools* due to natural forces.

These natural forces include variations in solar intensity (including sunspot cycles), cycles of warming and cooling of the oceans (El Niños and La Niñas are a product of a natural oscillation cycle in the Pacific Ocean, for example), astronomical factors (the Milankovitch Cycles, cosmic rays), tectonic shifts, volcanoes (including undersea volcanos), and so on.

These factors are dynamic and highly interactive; for example, if the Atlantic and Pacific Oceans are both in a "warming" cycle, then there will be considerable "global warming." If they are out of phase, we could get warming or cooling depending on the strength of each cycle.

In any case, these oceanic oscillations of climate are temporary and beyond our control.

Even the IPCC has admitted that the factors causing climate are too complex to be predicted in the long-term—and then it makes long-term predictions anyway.

'Human' causes of warming more due to cities and agriculture than CO_2

Is some of the current warming human-caused? Yes. Human activities have contributed to warming due, initially, to rising levels of CO_2 (up to, again, I'd suggest, about 300 ppm), but more importantly due to larger cities and expansion of agriculture, especially as the world population rises to eight billion.

The world's cities are growing almost exponentially to house humanity's eight billion people, increasing the "urban heat island effect." Agriculture, another warming factor, has also increased to feed this eight billion.

To "stop climate change," as the alarmists wish, we would have to *drastically* downsize cities and *drastically* reduce agriculture—and this drastic "de-growth" *is* the alarmist policy, as we saw in Chapters 13-16, regardless of how many millions (or billions) of people are harmed.

To the climate realist, the warming effects of cities and farming are just factors we will have to adapt to if we wish to continue Western-style civilization (keeping in mind that many alarmists do *not* want Western-style civilization to continue).

Carbon dioxide is a fertilizer, not a 'pollutant,' and is 'greening' the planet

Carbon dioxide is a fertilizer, not a "pollutant" (as the U.S. Environmental Protection Agency labels it). Increased levels of CO_2 have not devastated the planet, as the alarmists would have us believe, but are "greening" the planet, as we saw in Chapter 8.

Farmers would not aim for CO_2 levels of 1,000 ppm or more in their greenhouses if CO_2 was a pollutant rather than a fertilizer. More CO_2

also makes plants more drought-resistant. More CO_2 is plus for the planet's plant life, and therefore for humans.

IPCC 'science' based on 'consensus' rather than facts

"The Science" behind global-warming alarmism is not "traditional," empirically based science, although the alarmists would like the public to believe this is the case ("The Science is settled"). As Thomas Sowell put it, the IPCC and its supporters are presenting "the results of essentially consensual processes ... as scientific."[4]

Yes, of course, there is empirical evidence underlying the many research articles that go into the IPCC reports. But this data is not handled, as in traditional science, with rigorous experiments and risky predictions that "stress-test" the data.

Instead, the IPCC, and all of alarmist climate science, rests on "post-normal" science that, as we've seen, relies on "consensus" rather than empirical evidence as its basis. This "consensus" approach offers a lot of leeway for subjective opinion (a.k.a., bias/ideology).

IPCC 'science' is highly influenced by political ideology—'review by governments'

The IPCC's "science" is highly influenced by political interference—for some reason the IPCC believes "review by governments" is an "essential" part of its process.

Although "review by governments" may also, and undoubtedly does, influence the processes of other, more empirically based sciences, these other sciences are not so blatantly in the pocket of politicians. Science that is "reviewed by governments" is not reliable science. Period.

As part of this political influence, the IPCC's important Summaries for Policymakers (important because most people only read the summaries rather than the full reports) are crafted in the main by a small group of environmental activists, with input from the government representatives.

This process is hardly the best guarantee of the "objectivity" and "transparency" promised by the IPCC mission statement, which is why the summaries, and the publicity that follows them, are consistently more alarmist than the reports themselves.

Climate alarmists happy to lie and/or distort the facts

It is a fact, shown over and over in this book, that climate alarmists are willing to lie and/or distort the facts to promote their ideology.

For example, Al Gore's *An Inconvenient Truth* strongly implies that CO2 determines temperature in the film's centrepiece wall chart. In fact, as even alarmist climate scientists acknowledge, temperature change *preceded* CO2 change over the 650,000 years of the chart, not the reverse as Gore strongly implies.

Yet Gore's movie was enthusiastically received by many climate scientists, some of whom actually praised the film for its accuracy (thereby stretching the definition of "accuracy" virtually to infinity) because it has the right "message." Science that accepts inaccurate facts if the "facts" give the right "message" is corrupted science.

Stephen Schneider acknowledged that alarmists had to offer "scary stories" to get public attention. The venerable David Attenborough's nature documentaries aren't so much inaccurate as rife with one-sided arguments and examples to make the "global-warming" situation seem much worse than it actually is by any objective standard.

And so on, as detailed in Chapter 12 and elsewhere. If there is another scientific discipline promoting such blatant dishonesty so consistently, I am not aware of it.

Even IPCC no longer promotes 'extreme' warming scenarios

The latest IPCC reports (especially AR6) no longer promote the more "extreme" future climate scenarios (RCP8.5/SSP5-8.5). That said, the Summaries for Policymakers *do* promote these extreme scenarios, which are then trumpeted by the UN leaders and the media to give the impression of an immanent dire "emergency."

And, incidentally, as we saw in **Sidebar 6: Radiative forcing**, radiative forcing of 8.5 Watts per square metre (as in RCP8.5 and SSP5-8.5) cannot produce 4.5°C of warming, as the IPCC claims, at least if we accept to the Stefan-Boltzmann Law (which most physicists do). At most, at least in Reality World as opposed to Climate Wonderland

World, radiative forcing of 8.5 W/m^2 can create about 1.5°C of warming.

Once again, climate alarmists are forced to suspend the laws of physics to accommodate a flawed theory.

Climate science is far from 'certain'

"The Science" is not "certain" and "settled," and therefore beyond criticism (and certainly not from the public!), as climate alarmists like to claim. For a start, as physicist Richard Feynman put it, "a scientist is never certain."

But, in their better moments, alarmist climate scientists themselves admit that their discipline is perhaps not quite as "settled" as they would like the public to believe. For example, as we saw in Chapter 10, ultra-alarmist Stephen Schneider noted there is a "maddening degree of uncertainty" in climate science. Then climate alarmism kicks in and we return to square one in terms of honesty with the public (i.e., back to dishonesty).

Net Zero will be absurdly expensive for little gain

Net Zero by 2050 (or by any year, really) will be ridiculously expensive (possibly \$275-trillion, or \$9.5-trillion a year) for almost no gain in net "cooling," as we saw in Chapter 14. If spending this amount of money will have almost no effect on temperature, why are we even considering it? Where is the cost-benefit analysis by governments? Does Bjorn Lomborg have to do *everything*?

Neo-Marxist agenda underlies climate alarmism

The answer to why "we are even considering" spending this vast sum of money for almost no climate benefit is because the United Nations, the IPCC, and other supra-national organizations (looking at you again, World Economic Forum) have another agenda.

They sugar-coat their agenda as making the world economy more "sustainable," or as saving us from ourselves (we're consuming so much we'll eventually "eat up" the whole planet!).

The real agenda, as I've tried to show in Chapters 15 and 16, is to radically restructure our social, political and economic systems along what can only be called communist lines, in the process destroying liberal Western civilization as we know it.

We'll all own less (and love it!), we'll consume less, we'll live in smaller, more self-sufficient communities (similar to Benedictine monasteries), we won't be "enslaved" by technology, we'll be in harmony with Nature ("reenchanted"). In short, the not-so-hidden climate-alarmist agenda is a form of utopian social engineering.

I say "not-so-hidden" because the alarmist "de-development," anti-growth, radically socialist agenda is actually right out in the open (UNEP's "Making Peace with Nature," for example). But the project is so insane that few in the public, politics or media actually believe this idealistic future Utopia could be the climate alarmist goal.

This utopian Net Zero social restructuring *is* the alarmist goal and they are open about it. We should take their words seriously.

Climate alarmists aim for 'inner transition'—i.e., brainwashing

Just as climate alarmists wish to radically restructure society, they also want to radically restructure our brains (the "inner transition") so that we all, as it were, learn to love the Green Big Brother.

The benighted people who don't want to be reprogrammed in this way are regarded as archaic holdovers from the Old Regime (including "deniers") and will have to be gently (really?) persuaded to accept the Brave New Green World.

This ideal communist society has never worked before (Soviet Russia and Communist China have murdered millions in the name of Marxist idealism). For some reason—perhaps because human nature is not as plastic as the idealists would like—people tend to resist, at least internally, this kind of totalitarianism.

Under the alarmist program, which includes the elimination of "liberty and individualism," those hold-outs will be brought to see the error of their ways. Unless, of course, those of us not in the thrall of alarmist ideology fight back....

Finding a 'true view' through *real* science

The above is only a partial listing of the flaws, distortions and problems with alarmist climate science, as embodied by the IPCC and its supporters, that I've presented in this book.

Alarmist climate science claims, at least to the public, to be a "hard" science with strong, empirical evidence.

However, I don't believe a diligent researcher could find a similar pattern of corruption of science and blatant political ideology in any other discipline in the "hard" sciences. (The social sciences are much more prone to this kind of ideological manipulation.)

To repeat the question of Jean-François Revel: "How is it possible for a theory, which is false in its component parts, to be true as a whole?"[5]

I've several times cited Aristotle's dictum that "with a true view all the data harmonize, but with a false one the facts soon clash."[6]

Apart from having parts that are out-and-out "false," alarmist climate theory also doesn't "harmonize" as a whole in the Aristotelian sense, which strongly suggests it is not a "true view."

For example, alarmist climate theorists are quite capable of believing two *completely contradictory facts* at the same time, echoing the "six impossible things" that Alice's White Queen can believe before breakfast, or Orwell's doublethink: "To hold simultaneously two opinions which cancelled out, knowing them to be contradictory and believing in both of them."[7]

Here are four "impossible things" (or at least contradictory things) that climate alarmists believe:

1. Climate alarmists recognize we are in an Ice Age, geologically, with temperatures the coldest in at least 275 million years. Yet alarmists fear a degree or two Celsius of warming.
2. As we saw in Chapter 8, climate alarmists can, on the one hand, warn of extreme warming ahead due to the evil carbon dioxide, while at the same time labelling the *much warmer* periods of several millions of years ago, with *much higher* CO_2 levels, "a tropical paradise."

3. Climate alarmists believe that adding carbon dioxide to the atmosphere will lead us to a dead, burned-out planet, while at the same time being fully aware that this added carbon dioxide is a plant fertilizer that is actually "greening" the Earth (also noted in Chapter 8).
4. As we saw in Chapter 13, climate alarmists believe that "climate change" will devastate human civilization if we don't drastically curb CO2 emissions, while at the same time noting that the "business as usual" scenarios—SSP5/RCP8.5—will increase GDP by up to ten times by 2100. How can a devastated planet increase our GDP?

If these examples of doublethink don't show a lack of "harmonization" in alarmist climate theory, I don't know what would.

In more modern terms, alarmist climate theory is incoherent. Incoherent theory is not good theory, and certainly should not be the basis for major political and economic decisions. What we need, if we actually need to "combat climate change," is *good* theory that is internally consistent.

Western civilization rests on science and reason

This is why we have science—indeed, liberal Western civilization rests on science (a negative attainment for alarmists like Al Gore and Morris Berman since it "separates us from Nature"). It is science that allows us to distinguish the real from the imaginary, the true from the false.

In other words, at least ideally, Western civilization is based in reason rather than emotion ("I *feel* it is true, therefore it *is* true"), on evidence rather than "consensus." And Western-style civilization has been spectacularly successful, for all its faults and flaws, in improving the lives of billions of people.

It was Western science that saved humanity from the "Malthusian trap" of growing populations that inevitably run out of arable land, leading to mass starvation and a subsequent population crash. We may not be "enchanted," as the alarmists wish, but thanks to Western medicine, if nothing else, billions of us are alive today who would otherwise never have existed.

Neo-Malthusians aim for 'de-growth'

Yet today we face a new crop of Malthusians, the neo-Malthusian climate alarmists, and environmental alarmists generally, with their claims that we are still in Malthus's trap and will therefore face catastrophe if we don't follow their Net Zero "de-growth" program—a program that will itself, as I've argued, be far more catastrophic for humanity than a degree or two or even three of "global warming."

The utopian social engineering of Net Zero will take us back to a form of the Middle Ages, making us all poorer (except for the privileged elite) by spending trillions of dollars that will have close to zero effect on the climate. This utopian, "tear-it-down" approach is, in the context of climate, as insane as any Mad Hatter's tea party.

In particular, to *artificially* make the energy source most used by modern society more expensive, based on the shaky claims of "post-normal" science, is a recipe for far greater disaster than anything the climate can throw at us. Net Zero is insane.

Yes, at some point in the future fossil fuels will become scarcer and more expensive and we will have to find alternatives, assuming we want a liberal Western-style civilization to continue (which most climate alarmists do *not* want, by the way, as we've seen; the average person may feel differently).

But if we can get out of the "apocalyptic global-warming," "sustainable-energy," "Net-Zero-by-2050" straitjacket we will be much freer and give ourselves much more time to invest in energy alternatives—which means, I think, nuclear power—that can actually support prosperity rather than destroying it.

"Sustainable" wind and solar may be part of the mix, but only a small part. Unless, of course, we really do want to reduce our standard of living to that of a "very poor third world country," living in small Benedictine-monk-like communities based on cottage industries, as suggested by Andrew Nikiforuk in *The Energy of Slaves* and Morris Berman in *The Reenchantment of the World*, both channeling Karl Marx.

I don't see how the de-development visions of the utopian alarmists can, realistically, ever occur without totalitarian compulsion and hundreds of millions if not billions of deaths.

The lives of most of the current eight billion humans are highly dependent on the fossil fuels and artificial fertilizers that the alarmists hope to eliminate.

For example, artificial fertilizers are one source of the Green Revolution that has prevented the mass starvation predicted by Paul Ehrlich's ultra-alarmist *The Population Bomb*.

And once we have accepted a totalitarian society, we may find it very difficult to return to anything resembling democracy.

Democracy vs. totalitarianism

Even as I write this, I can see choices being removed from average people, without their consent or permission, by government fiat.

As mentioned, no more sales of cars with internal-combustion engines allowed by 2035 for Canada. No more construction of buildings using natural gas (a regulation already passed in Victoria and pending in other cities). No more natural gas at all, as the City of Ottawa wishes, should its natural-gas pipeline fail.

What other restrictions (hardships?) can we reasonably foresee in the low-carbon future planned for us? Airplanes emit a lot of carbon, so probably flying will be forbidden for the average person (but not, of course, for the elite attending their international climate conferences)—why would a Benedictine monk, happy in his/her small community, want to fly, anyway? Why, indeed, would a Benedictine monk want or need "liberty and individualism"?

Under Net Zero we are facing massive changes to our society, moving us toward what can only be called a neo-Marxist model under the guise of "environmentalism" and "fighting climate change."

But we are not being told by our governments, plainly and publicly, the sacrifices we will be expected to make—because if we were told, most of us would reject the Net Zero option out of hand and demand other, less radical options.

A government that uses a politically manipulated scare story (the "climate crisis") to dictate to its citizens what they are allowed to eat, what kinds of cars they must drive, how they must heat their homes, the kinds of occupations they are allowed to follow (after a "just transition"), that *deliberately* lowers its people's standard of living, that expects an "inner transition" in our thinking and behavior to make all this utopian social engineering work (we'll all come to love Big Brother), is not a democratic government. It is a totalitarian government.

At the time or writing, the alarmists are winning the propaganda battle for this Brave New Society: Marxist true believers and fellow travellers increasingly control the levers of power and information in governments, universities and even primary and secondary schools (often under the banner of "wokism" and its offshoots).

This should worry us; Marxist utopianism is an ideology we should resist with every power at our command if we wish to retain at least some of the freedoms that we, in democratic nations at least, have enjoyed for the past two centuries or so.

And once freedom is lost, as I've said several times because I think it's important, getting freedom back will be very difficult. Just ask the citizens of Russia, who appeared to be on a path to a freer society after the fall of the Soviet Union in 1991 but have lapsed into a form of strong-man, expansionist Stalinism under Vladimir Putin.

Because we in the democratic countries have been free so long, without a major war in 75 years, we have taken our current peace and freedom for granted, rather than as something we need to continually fight for.

I say without qualification that the current scare over "global warming"—the so-called "climate crisis"—is based far more in political ideology than actual climatic conditions. We are being manipulated by an elite, international group (I'm looking at you, United Nations and World Economic Forum, among others) that wishes to reduce our lifestyles and remove our freedoms under the banner of "sustainability" and "saving the planet."

Making a choice

At this point in history, we as citizens of the world need to make a choice.

Are we willing to sacrifice a comfortable material lifestyle in the, I guarantee, futile quest to somehow control the climate when we can't even control the weather? In that case, we have chosen the *alarmist* side of the looking glass, and fair enough, as long as we're fully aware of what we're choosing.

Or are we as citizens willing to say "no" to the Net Zero fantasy in favor of an *improving* standard of living for ourselves and our children, a standard of living that will allow us to handle whatever climate changes *actually* occur (as opposed to the bogeymen in the climate models and alarmist propaganda)? In that case, we have chosen the *realist* side of the looking glass.

The return to sanity

If we want to fight back against Net Zero, where do we begin?

We begin, I think, by demanding from our governments *facts and figures* about what Net Zero by 2050 will *actually* cost us—in other words, a basic cost-benefit analysis. So far, they have not done so beyond vague promises of a "better" future (e.g., a "just transition") once fossil fuels have been wrestled back into the ground. Sure.

I don't think we'll ever get this full disclosure, by the way. A full accounting of the costs and benefits of "fighting climate change" is too damaging to the alarmist cause.

So perhaps our next step is to support and elect political parties that are not captivated by the Net Zero fantasy and willing to deal with "climate change" rationally and realistically.

There is a problem here. The opinion polls tell us that many if not most people believe, as good environmentalists, in the alarmist social-engineering program (as noted in Chapter 5, the alarmists have been having a public-relations field day with the "record" temperatures of summer 2023). Some people even think they face "extinction" if the planet warms a degree or two.

In that case, a reason-based political party must have the courage to *lead* public opinion against the alarmist program. A good start would be a policy of eliminating the "carbon tax" or equivalent, a completely unnecessary expense that makes energy more costly while having virtually zero impact on climate.

Another good policy: abandon Net Zero by 2050 in favor of a much longer timeline, looking ahead to the time when, yes, fossil fuels will be scarcer and much more expensive and will need to be replaced. But until then we should be using the fossil-fuel-based economy to finance this transition to, almost certainly, some form of nuclear power.

We should formally demand a "piecemeal engineering" approach to the problems that climate does throw at us rather than alarmist utopian social engineering. For thousands of years human beings have coped with fires, droughts, floods, extreme weather, even sea-level rise. Why would we not continue to cope today when we have so many sophisticated technological tools at our disposal?

Similarly, we should demand that social and economic policies be based on *real* science, not "post-normal" science. At the very least, we should regard "post-normal" science with a great deal of skepticism—in this book we've seen many examples of how the IPCC and its cheerleaders have distorted "The Science" to fit a preconceived political ideology.

Climate alarmism is 'stealing our future'

Greta Thunberg, the poster child (literally) for climate alarmism, laments that climate change is "stealing my future." On the contrary, it is climate alarmism, if allowed to continue, that really will steal the future of our kids and grandkids, making them both less prosperous and less free.

We can continue on our current path down the rabbit hole of Net Zero by 2050 and reduce our children and grandchildren to living in the equivalent of "a poor third-world nation."

Or we can abandon this manufactured and irrational fear of "global warming" and simply resolve to adapt to "climate change," an approach that will give our children and grandchildren a freer, more prosperous life. Which side of Alice's mirror do we want to be on?

I hope we make the right choice and reject climate alarmism in favor of rational, realistic adaptation to "climate change" as part of the improving global society that could be ours.

But for this rational and realistic adaptation to occur, we must be willing to fight as if our futures, and our children's futures, depend on winning this battle. Because they do.

Notes to Chapter 18

---

1 For example, see "Climate 'Realism' is the New Climate Denial," by Stella Levantesi and Guilio Corsi, *The New Republic*, Aug. 6, 2020.

2 Damien Carrington, "The four types of climate denier and why you should ignore them all". *The Guardian*, July 20, 2020.

3 Many of these accomplished people are, like me, members of Climate Realists of British Columbia. Our website is climaterealists.ca. If you visit there you will find a wealth of articles promoting a rational response to "climate change," with plenty of facts. But also, have a look at the "Members" section and you'll see I'm not exaggerating about our members' professional qualifications.

4 Thomas Sowell, *Knowledge and Decisions*. New York: Basic Books, 1980 (1996 edition), pp. 5-6.

5 Jean-François Revel, *Neither Marx nor Jesus*. New York: Laurel, 1971, p. 15.

6 Aristotle, *Ethics*, Book I, Section 8.

7 George Orwell, *1984*. Toronto: Penguin, 2008 (1949), p. 37.

About the author

For almost 40 years I was a journalist (editor and writer) for a number of news organizations, including *The Canadian Press, The Toronto Star, The Globe and Mail, The Bangkok Post,* and most recently the Victoria *Times Colonist,* where I was an editorial writer and weekly columnist. In 2002 I left the *Times Colonist* to enroll in the University of Victoria graduate program in English. I got my MA degree in 2005.

From 2005 to 2020 I taught writing in various formats, mostly business and professional writing, for the University of Victoria, University Canada West, Royal Roads University, and the online Athabasca University. (For those who might wonder, I have never gotten any money from the oil companies!)

As an ex-journalist writing on science, one of my inspirations is Canadian investigative journalist Elaine Dewar, a non-scientist who is not afraid to tackle scientific issues.

Her excellent 2001 book *Bones: Discovering the First Americans* probes the controversy in paleo-anthropology over how North America's first peoples came here: was it via a land-bridge across Alaska about 12,000 years ago or by sea thousands of years earlier?

Anthropologists who believed the by-sea theory were, Dewar reports, treated much the same as climate skeptics: they were "heretics" who were academically shunned and suppressed. That said, the "earlier arrival by sea" theory is more accepted today by paleo-anthropologists.

On her credentials, Dewar writes:

> So why, you might ask, should a journalist investigate a question that has bedeviled scientists for 150 years? Journalists are willing to go anywhere, to be passionate fools, to ask innocent questions, to ignore barriers, to look for

> patterns that connect disciplines and solitudes, and to have no vested interest in anybody's intellectual capital.
>
> Science must be public and transparent or it loses all meaning, and journalists bring evidence out of the lab and spread it before the general public. We are also a little like bees. We dip into everybody's business and carry the news along. We cross-pollinate. We fertilize. Sometimes we sting.[1]

As with my first book, *False Alarm,* I hope *Through the Looking Glass* also cross-pollinates, fertilizes, and stings.

---

[1] Elaine Dewar. *Bones: Discovering the First Americans.* New York: Carrol & Graf, 2001, p. 6.

Made in the USA
Monee, IL
04 November 2026

67387182R00144